Nuffield Design & Technology

Product
design

Addison Wesley Longman

Edinburgh Gate, Harlow, Essex, CM20 2JE, England, and Associated Companies throughout the World

First published 1996
ISBN 0582 23469 7

Set in ITCKabel and Times
Designed and produced by Pentacor, High Wycombe, Bucks HP12 3DJ
Illustrations by Nathan Barlex, Oxford Illustrators, Pentacor
Picture researcher Louise Edgeworth
Editor Katie Chester
Indexer Richard Raper/Indexing Specialists
Printed in Great Britain by Scotprint Limited, Musselburgh, Scotland

The publisher's policy is to use paper manufactured from sustainable forests.

Project Directors
Executive Director Dr David Barlex
Co-directors Prof. Paul Black and Prof. Geoffrey Harrison
Deputy director – Dissemination David Wise

Contributors

David Barlex	Michelle Bell	Nick Givens
India Hart	Terry Hewitt	Deborah Howard
Glyn Jones	Paul Laurie	John Plater
Judith Powling	Ann Riggs	Ian Steel
David Wise	Dorothy Wood	

We are grateful to the following for permission to reproduce photographs and other copyright material:

Jane Adam 51A (Crafts Council); Art Directors Photo Library/Terry Why 106; BBC Copyright © 29; John Bigelow Taylor Photography 136; Gareth Boden 46, 47, 56B, 59, 70, 126, 127, 209; British Museum 48; Cabaret Mechanical Theatre 114B, 115AR,115B & 117 (Heini Schneebeli), 115AL (Gary Alexander); Crafts Council 86L (A.M Schillito), 86R (Ian Dobbie); Mary Evans Picture Library 28, 32, 36A; Anne Finlay 51B (Crafts Council); Fletcher Priest/Ian McKinnell 26, 27R; Robert Harding Picture Library/Jon Gardey 58; Irwin Desman Ltd 43, 44, 45; The London Planetarium 27L; London Transport Museum 36B, 37, 38; Longman Photographic Unit 31B; Mattel UK 59A; Paul Mulcahy 208; John Plater 104, 112, 188, 191, 192, 193; Popperfoto 30; Powergen plc 25; Press Association/Martin Keene 31A; Quart de Poil 62; The Royal Aeronautical Society 35; The Science Museum/Science & Society Picture Library 52, 57; Science Photo Library 56A (Adam Hart-Davis); Tony Stone Images 8 & 9 (Ian Shaw), 100 (Pete McArthur); Telegraph Colour Library 19 & 20 (Benelux Press), 40 (Space Frontiers); Thorn Lighting Ltd 54; Timber Kits 114A; Unilab 15, 89; Unipath Ltd 33; Grant Walker 50 (Gareth Boden).

The Nuffield Design and Technology Project gratefully acknowledges the support of the following commercial concerns in developing the published materials:

ApproTEC	Fletcher Priest	Irwin Desman Ltd.
Grant Walker Ltd.	Maclaren	Rexam Plc.
Unilab		

Contents

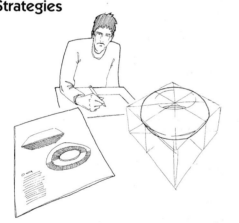

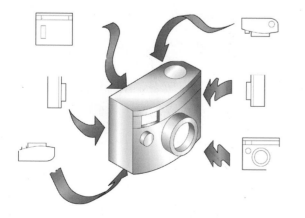

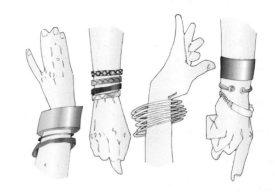

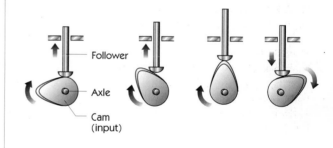

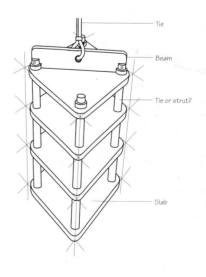

Contents

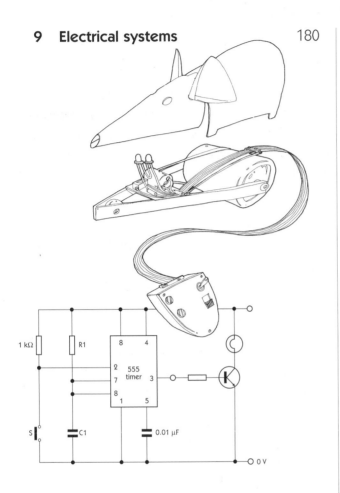

Part 1 Learning D&T at 14–16

What will I design and make?

During key stage 3 you used a wide range of different materials for designing and making – textiles, food, wood, metal and plastic. At key stage 4 you are allowed to specialize in a materials area. You have chosen to specialize in designing and making with *resistant* materials. This means that you will be using wood, metal, plastic or a combination of these materials. You may also use components to produce mechanisms or electrical circuits.

The reason for specialization is that at key stage 4 you are expected to work to a higher standard in both designing and making; the quality of your products should be better than at key stage 3. The key stage 4 course lasts only two years and you simply don't have enough time to gain the extra skills, knowledge and understanding needed to improve your work in more than one materials area. The sorts of things that you will be designing and making are shown below.

This area of designing and making is usually called **product design**.

Of course there is more to design and technology than designing and making and in your key stage 4 course you will also learn about the way design and technology works in the world outside school. In particular you will study how industry is organized to manufacture goods.

▶ *Designing and making at key stage 4 is a real challenge. Your products should be good enough for the shops*

How will I learn?

If you do design and technology the Nuffield way then your teacher will use three different teaching methods. These are described below.

Resource Tasks

These are short practical activities. You should find that they make you think and help you learn the knowledge and skills you need to design and make really well.

Case Studies

These describe real examples of design and technology in the world outside school. By reading them you find out far more than you can through designing and making alone. Case Studies help you to learn about the ways firms and businesses design and manufacture goods and how those goods are marketed and sold. You will also learn about the impact that products have on the people who use them and the places where they are made.

Capability Tasks

These involve designing and making a product that works. When you tackle a Capability Task you use what you learn through doing Resource Tasks and Case Studies. Capability Tasks take a lot longer than either Resource Tasks or Case Studies. Your teacher will organize your lessons so that you do the Resource Tasks and Case Studies you need for a Capability Task as part of the Capability Task. In this way your teacher makes sure that you can be successful in your designing and making.

The way these methods work together is shown here.

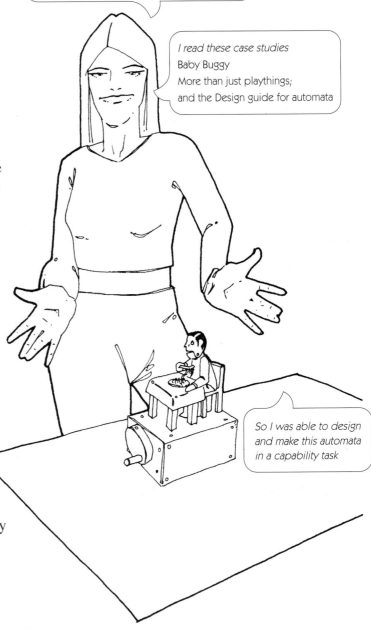

I did these resource tasks:

SRT3   Design Briefs and Specifications
SRT4   Brainstorming
SRT5   Attribute analysis
SRT7   Systems and control
CRT2   Communicating ideas to the maker
MRT2   Design Challenges 2 – A walking toy
MRT3   Design Challenges 3 – Show and turn
LIRT   Automata – Making a dragon

I read these case studies
Baby Buggy
More than just playthings;
and the Design guide for automata

So I was able to design and make this automata in a capability task

Resource Tasks for gaining knowledge, skills and understanding

You will be given a Resource Task as an instruction sheet like the one below. All Resource Tasks are laid out in the same way. You will see that they are different from the ones you used at key stage 3.

code number   title

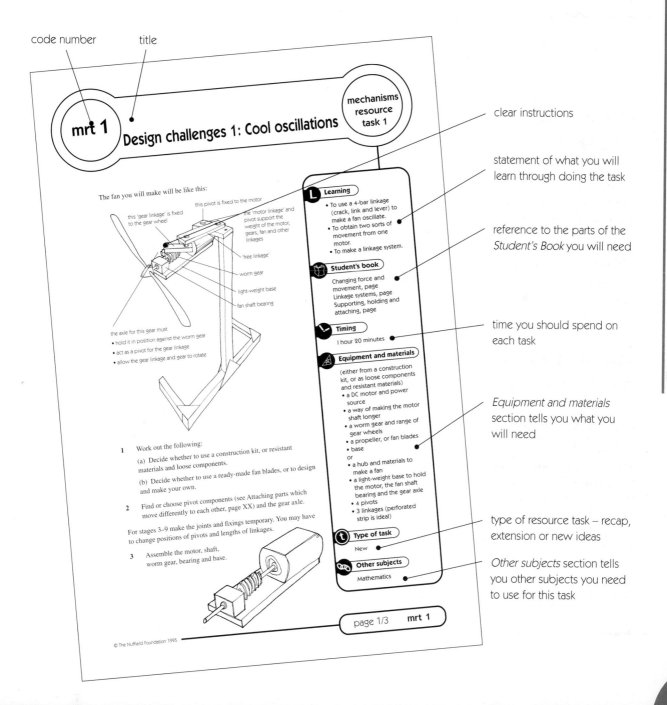

clear instructions

statement of what you will learn through doing the task

reference to the parts of the *Student's Book* you will need

time you should spend on each task

Equipment and materials section tells you what you will need

type of resource task – recap, extension or new ideas

Other subjects section tells you other subjects you need to use for this task

More about Resource Tasks

There are three types of Resource Task.

ALL STUDENTS - SRT1 RECAP (MON)
TABLES 1 2 + 3 SRT2 - EXT (WED)
TABLES 4 5 + 6 SRT3 EXT (WED)
TABLES 1 2 + 3 SRT3 EXT (MON)
TABLES 4 5 + 6 SRT3 EXT (MON)
ALL STUDENTS SRT4 NEW IDEAS (WED)

Your teacher may introduce a sequence of Resource Tasks by talking to the whole class

Recapitulation Resource Tasks

These are tasks that go over things that you probably did during key stage 3. They are very useful for reminding you of things you may have forgotten about or for catching up on things you have missed.

Extension Resource Tasks

These are tasks that take an idea that you were probably taught at key stage 3 and develop it further. They are useful for both revising key stage 3 ideas and helping you use them in a more advanced way.

New ideas Resource Tasks

These are tasks that deal with knowledge and understanding that are new at key stage 4. It is unlikely that you will have done this work at key stage 3. They are important for helping you to progress.

Your teacher may:

- organize the lesson so that everyone is doing the same Resource Task;
- set different students different tasks;
- allow you to choose from a range of Resource Tasks.

Sometimes you will work on your own and sometimes as part of a team.

Case Studies for awareness and insight

There are two types of Case Studies at key stage 4.

The first type are those that deal with 'large' technologies. These are the technologies which significantly affect the way people live. Often they are associated with a particular period in history. It is important that you read these Case Studies because they will help you to understand the way that technology affects our lives.

The second type are those that deal with products that are similar to those that you will be designing and making yourself. They describe:

● how the designs were developed, manufactured, marketed and sold;

● how the products work;

● how the products affect the people who make them, those who use them and others.

A particular Case Study may deal with just one of these or with all of them. It is important that you read these Case Studies because they will help you to design like a professional designer.

It is easy to lose concentration when you are reading a Case Study so they all contain Pauses for thought and Questions which you should try and answer while you are reading them. It is often useful to discuss your answers with a friend. This will help you both to think about and make sense of the study.

The Case Studies also contain Research activities. You will often be set these for homework as they involve finding out information that is not in the Case Study. This is important as it will help you to learn how to get new information as well as understand more about design and technology.

◗ *A student presents his Case Study research findings to the whole class*

Capability Tasks for designing and making

Each of the products you design and make at key stage 4 will be from a group of product types. These groups of product types are called **lines of interest**. So, for example, you might design and make a product that was from the line of interest 'seating'. Your product could range from a small, portable, three-legged collapsible fishing stool to a large bench designed for use in a city centre. Certain sorts of knowledge, skills and understanding are useful for designing seating, whatever type of seating it might be. These include an understanding of structures, the properties of materials, construction techniques, ergonomics and anthropometrics and aesthetics – all these are needed to design and make good seating.

We have suggested seven lines of interest for Capability Tasks in the area of product design. Some possible products from some lines of interest are shown here.

Seating

Storage

Automata

Games, toys and playthings

Lighting

Testing equipment

Body adornment

During your key stage 4 course you will have the opportunity to work in at least three different lines of interest. If you were to work in only one line of interest, while you would end up knowing a lot about that particular part of design and technology, there would be other parts you would know nothing about at all. If you were to work in many more than three lines of interest you wouldn't have the time to study anything in depth, so you would end up knowing very little about any part of design and technology. So working in three lines of interest will enable you to gain a reasonable level and range of knowledge, understanding and skill in design and technology.

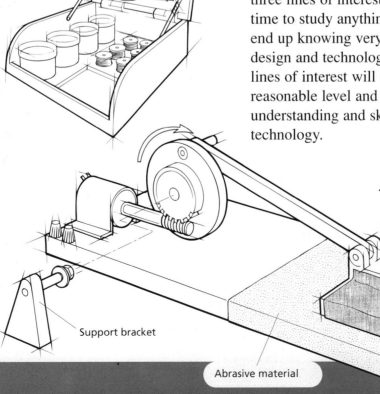

Textile sample

Support bracket

Abrasive material

Managing three Capability Tasks

If you are following a full GCSE course, it is likely that you will tackle three Capability Tasks during year 10, each one from a different line of interest. Your teacher will work out with you which ones your class will tackle. In year 11 you can either revisit a line of interest or tackle a new one. The one in year 11 will probably be used for your GCSE coursework. This makes sense because you should be better at designing and making in year 11 than you are in year 10.

It will be quite a struggle to fit three complete Capability Tasks into year 10 so your teacher may organize the lessons so that you only do part of some of these tasks. You will certainly need to do one complete Capability Task where you design, make and test a well-finished product. In another Capability Task you might only produce a working model or collection of models of the product. This means that you don't have to spend a lot of time making the finished article. In another Capability Task you might only produce a series of design proposals as detailed annotated sketches. This cuts down the time you spend on the Capability Task even further.

Your teacher may give your class a design brief plus a specification and ask you to design and make a product that meets those requirements. Your teacher might even give you the brief, the specification and the working drawings and ask you to make the product so that you can learn about the manufacturing process. Of course it is important that you carry out the Resource Tasks and Case Studies needed for each of these Capability Tasks. In this way you acquire a lot of design and technology knowledge, understanding and skills and still keep in touch with designing and making. This will put you in a strong position to tackle a full Capability Task in year 11.

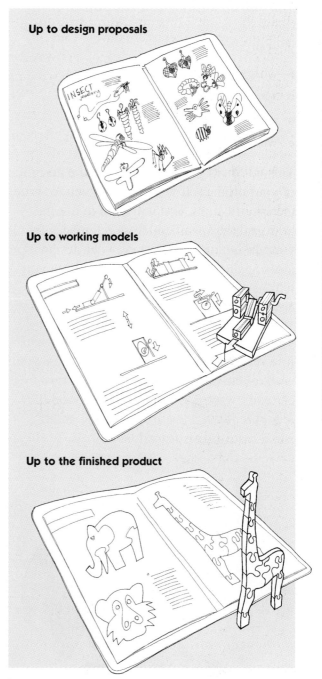

Up to design proposals

Up to working models

Up to the finished product

A Capability Task can be work completed at different stages

Ensuring your designing makes sense

You will be working to a brief which summarizes the following information about your product:

- what it will be used for;
- who will use it;
- where it might be used;
- where it might be sold.

This will help you to think about the design of your product. It will also help you to write the specification. You will need to use the brief and the specification as references for your designing. By checking your design ideas against the brief and specification you will be able to see whether they are developing in sensible directions.

This checking is often called **reviewing** and it is very important. If you fail to review your work at the correct times you will almost certainly waste a lot of time and your design ideas are likely to be inappropriate and in some cases may not work at all.

First review

Once you have some ideas for your product in the form of quickly drawn annotated sketches, then you should carry out your first review by comparing your ideas with the requirements of the brief and the specification.

Ask yourself the following questions for each design idea.

- Will the design do what it is supposed to?
- Will the design be suitable for the users?
- Will the design fit in with where it might be used or sold?
- Is the design likely to work?
- Does the design look right for the users and sellers?
- Have I noted any special requirements the design will need to meet later on?

Any design ideas that do not get a 'yes' to all these questions should be rejected or adjusted. In this way you can use the first review to screen out any design ideas that do not meet your requirements. You can do this screening in two ways:

- on your own, just thinking it through in your head and making notes against each design idea;
- working in a group and explaining your ideas to the other students in your group who can check them out against the questions. This takes longer and you have to help the others in the group to check out their design ideas as well. But the extra time is usually well spent as other people are often more rational in their criticism of your ideas than you are.

Whichever way you choose it will be important to discuss your review findings with your teacher.

Second review

By screening your early ideas you will be able to focus your efforts on developing a single design idea and on working out the details of that design. You will present these details as a mixture of annotated sketches, rendered presentation drawings and working drawings (sometimes called **plans**).

To make sure that your designing is still developing in a sensible direction you need to ask the following questions before you begin making the product.

- Am I sure that the working parts of the design will do what they are supposed to?
- Am I sure about the accuracy with which I need to make each part?
- How long will it take me to make and assemble all the parts of my design?
- Have I got enough time to do this?
- If not, what can I alter so that I have a design that I can make on time and that still meets the specification?
- Will the materials I need be available when I need them?
- Will the tools and equipment be available when I need them?
- Am I sure that I can get the final appearance that I need?
- Have I got enough time for finishing?
- Is there anything I can do to be more efficient?

You are probably the only one who can answer these questions but it will be worth checking your answers with your teachers as they are likely to know about any hidden traps and pitfalls.

Evaluating the final products

Here are examples of the ways in which you can evaluate your design once you have made it. You can find out more about them in the *Strategies* section (pages 68–95). It will be important to use all these different methods in coming to a judgement about the quality of your design.

User trip

By interviewing the user this student was able to find out what she did and didn't like about the brooch she had designed and made.

Performance specification

Gupta designed a test rig for looking at the strength of plastic curtain hooks. The specification for the equipment was as follows:

- to measure reliably and accurately the tensile force required to break a curtain hook;
- to record the results in a database;
- to enable 20 rings to be tested every hour.

He was surprised to find that he could only test 10 rings. It is important to design your product to meet all the specification requirements.

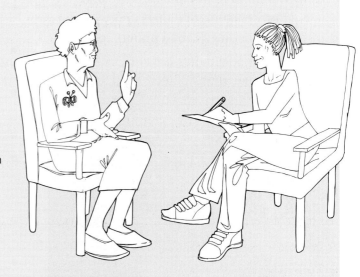

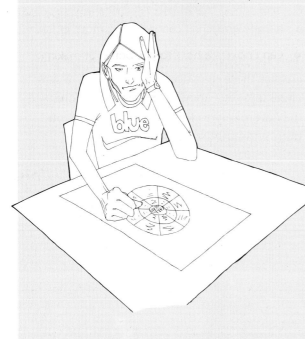

Winners and losers

Jane designed seating for listening to brass band concerts in a local park. It replaced the deckchairs which were wearing out and had to be put away each night. Jane's seats were fixed to the ground to prevent theft. It seemed at first that everyone would gain if her design was used. But what about using the seats for goal posts, arranging the seats for family groups, using the seats elsewhere in the park …? And what about the people who make deckchairs? By thinking about winners and losers Jane could see that it wasn't that simple.

Appropriateness

Fred designed a series of jigsaw puzzles for nursery schools to help children to learn about getting washed and dressed. The parts were simple to make and could be packaged easily. He hoped that they could be manufactured in a depressed area as part of a regeneration scheme. By asking the questions on page 93 he was able to decide whether his design was appropriate.

Thinking about how well your product meets its specification

One way to do this is to discuss your product with some other students. Give your product a star rating for each part of the specification – 5 stars if it meets that part really well, 3 stars if it meets it moderately well, 1 star if it meets it only poorly and no stars if it fails to meet this part of the specification. The next part is the tricky bit. Explain to the other students in the group why you have given the scores you have. Their job is to question your judgments. Your job is to convince them that the judgements are correct. If you do this you will be in a good position to move on to looking at your progress.

Looking at your own progress

At the end of a Capability Task it is important to look back at what you have done and reflect on your progress. The following sets of questions will help you with this.

Feeling good about what you have done

- Am I proud of what I made?
- Can I explain why?
- Am I proud of the design I developed?
- Can I explain why?

Understanding the problems

- What sorts of things slowed me down?
- Can I now see how to overcome these difficulties?
- What sorts of things made me nervous so that I didn't do as well as I know I can?
- Do I know where to get help now?
- What sorts of things did I do better than I expected?
- Was this due to luck or can I say that I'm getting better?
- Were there times when I concentrated on detail before I had the broad picture?
- Were there times when I didn't bother enough with detail?

- Can I now see how to get the level of detail right?

Understanding yourself

- Were there times when I lost interest?
- Can I now see how to get myself motivated?
- Were there times when I couldn't work out what to do next?
- Can I now see how to get better at making decisions?
- Were there times when I lost my sense of direction?
- Can I now see how to avoid this?

Understanding my design decisions

- With hindsight can I see where I made the right decisions?
- With hindsight can I see where I should have made different decisions?
- With hindsight can I see situations where I did the right thing?
- With hindsight can I see where I would do things differently if I did this again?

Part 2
Using other subjects in D&T

Using science

At key stage 4 you will be able to use science when you are tackling Capability Tasks. This is different from using science in a Resource Task. In a Resource Task you will be *told* to use science in the Other subjects section. In a Capability Task you have to *choose* when to use science.

Your science lessons will teach you two main things. First, how to carry out scientific investigations. If you need to find something out in a Capability Task, say the strength of a structure or material, the lifting force of a motor or the best lighting conditions for someone to read by, then you can use your science to help you plan the investigation and design the necessary experiments. Second, in science you will acquire scientific knowledge which could be useful to you in a Capability Task. The information in the table will help to remind you of the science you are likely to find useful. Note that some of the science is from key stages 2 and 3 as well as key stage 4.

Capability Task line of interest		Science likely to be useful
Seating		The properties and uses of materials (KS2); Springs and elastic bands (KS2); Skeletons, muscles and movement (KS3); Force and pressure (KS3); Stiffness and materials (KS4)
Storage		The properties and uses of materials (KS2); Stiffness of materials (KS4)
Automata		Springs and elastic bands (KS2); Simple circuits, switches and dimmers (KS2); Stiffness of materials (KS4); Velocity and acceleration (KS4); Power, current and voltage in circuits (KS4)
Games, toys & playthings		Simple circuits, switches and dimmers (KS2); Magnets, magnetic field and magnetic poles (KS3 & KS4); Power, current and voltage in circuits (KS4)
Lighting		Simple circuits, switches and dimmers (KS2); Formation of shadows, reflection and refraction and dispersion of light, effect of coloured filters (KS3); Power, current and voltage in circuits (KS4)
Testing equipment		Planning experimental procedures (KS3 & KS4); Obtaining evidence (KS3 & KS4); Analysing evidence and drawing conclusions (KS3 & KS4); Considering the strength of the evidence (KS3 & KS4)
Body adornment		Metals and non-metals (KS3); Chemical reactions (KS3); Electrolysis (KS4)

Uses of science in Capability Tasks

Using mathematics

At key stage 4 you will need to use your understanding of mathematics to help your design and technology. You will be able to use mathematics when you are tackling Capability Tasks. This is different from using mathematics in a Resource Task. In a Resource Task you will be *told* to use mathematics in the Other subjects section. In a Capability Task you have to *choose* when to use mathematics. Often you use mathematics without realizing it. The panel below shows some examples.

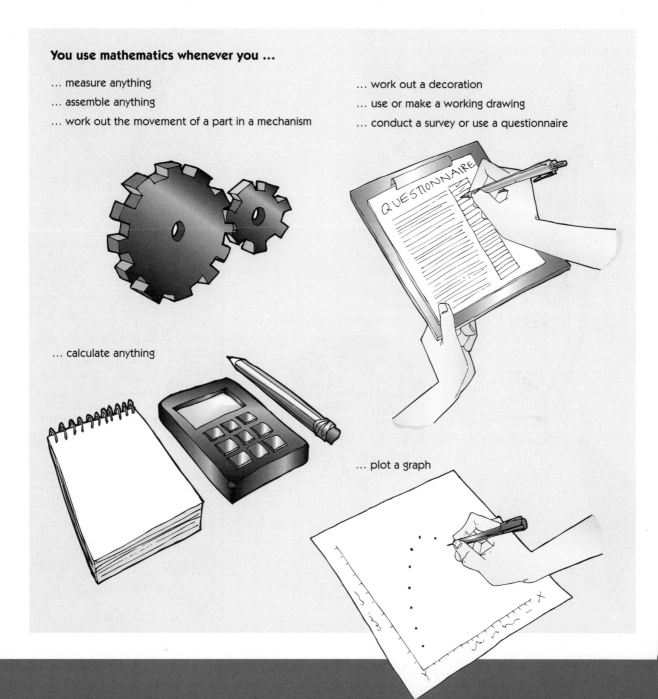

You use mathematics whenever you ...

... measure anything

... assemble anything

... work out the movement of a part in a mechanism

... work out a decoration

... use or make a working drawing

... conduct a survey or use a questionnaire

... calculate anything

... plot a graph

Using art

At key stage 4 you will need to use your understanding of art to help your design and technology. You will be able to use art when you are tackling Capability Tasks. This is different from using art in a Resource Task. In a Resource Task you will be *told* to use art in the Other subject section. In a Capability Task you have to *choose* when to use art. The example below shows how one student has used art in developing the design of a desk lamp.

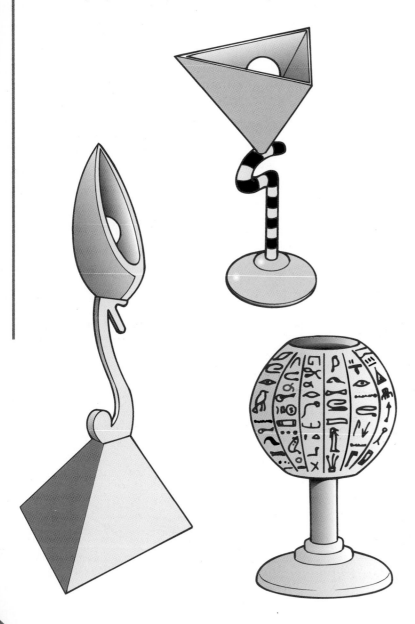

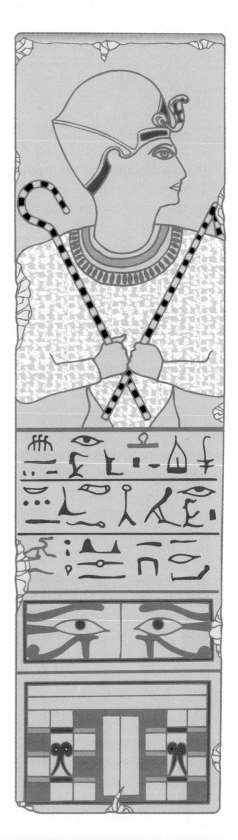

Using information technology

At key stage 4 you will need to use your understanding of information technology to help your design and technology. You will be able to use information technology when you are tackling Capability Tasks. This is different from using information technology in a Resource Task. In a Resource Task you will be *told* to use information technology in the Other subject section. In a Capability Task you have to *choose* when to use information technology. The examples below show how students used information technology to find out about existing products and to develop a reliable testing system.

TOY PURCHASE SURVEY
Write '1' in the box by each answer given, except where the printed instructions tell you to enter something else.

1 Have you just bought a toy?　'1' for yes ☐
　　　　　　　　　　　　　　'0' for no ☐

2 Is the toy for you?　'1' for 'yes', toy bought by a child ☐
　　　　　　　　　　'1' for 'yes' toy bought by an adult ☐

3 How old is the child for whom the toy was bought? ☐

4 Is the child a girl or a boy　'1' for a girl ☐
　　　　　　　　　　　　　　'1' for a boy ☐

5 Which one of the following best describes the toy? ☐
　A Construction & model kits　　B Board games
　C Soft & cuddly toys　　　　　D Dressing-up outfits
　E War & adventure toys　　　　F Figures & dolls: Sindy,
　　　　　　　　　　　　　　　　　　Barbie, Action Man etc.

6 How much did you pay? ☐

7 Had you decided what you would buy before you entered the shop, or did you look first and then decide?
　　　　　　　　Write '1' for decided before ☐
　　　　　　　　Write '1' for decided after ☐

8 Was your decision more based on what the toy was like or how much it cost?　Write '1' for what toy was like ☐
　　　　　　　　　　　　　　　　　Write '1' for price ☐

▶ *Questionnaire on toy purchase designed for use with a database*

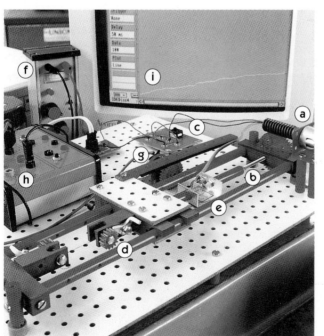

a The motor turns a large gear through a worm gear. Fastened to the centre of the gear is a threaded nut.

b As the gear turns the threaded rod is pushed or pulled.

c The direction the motor turns is controlled by a reversing switch. The circuit would be better if there was an on/off switch as well.

d The ratchet mechanism enables 'play' in the thread to be taken up without it unwinding when it is pulled.

e The strain gauge sensor changes resistance very slightly when the thread is pulled.

f The strain-gauge amplifier amplifies the very small change in voltage from the strain gauge into a noticeable output voltage.

g The displacement potentiometer is attached to a potential-divider circuit so that as the spindle is turned by the gear, the output voltage changes.

h The simple analogue interface converts the voltages from the strain gauge amplifier and the displacement potentiometer into appropriate signals for the computer.

i The software plots a graph on the screen of the force (pulling on the thread) against displacement (how much it stretches).

▶ *Test equipment linked to a computer for rapid, accurate data capture*

Part 3
How you will be assessed at GCSE
Writing your own Capability Task

It is likely that the Capability Task you tackle in year 11 will be the one that will be used for your GCSE coursework. This makes sense because you should be better at designing and making in year 11 than you are in year 10. Here are some guidelines to help you.

Designing the Capability Task

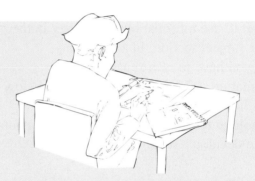

1 Deciding on the line of interest

Ask yourself these questions.

- Do you want to revisit a line of interest from year 10, or do you want to try something new?
- Which Resource Tasks did you enjoy most? Are these linked to a line of interest?
- Is there a group of students in your class who want to work on a particular line of interest?

2 Justifying your decision

Ask yourself these questions.

- Who will benefit from the product you are going to design and make?
- Will you be successful at designing and making this sort of product?
- Can you afford to make this sort of product?

3 Sorting out any extra learning that might be necessary

It is not difficult to identify particular areas of design and technology knowledge that are likely to be useful for your task. Discuss this with your teacher and identify Resource Tasks that could be useful.

4 Identifying any Case Studies that might provide useful background reading

Read and make notes listing those points that are relevant to your task.

5 Drawing up a 'Using other subjects' checklist

- Discuss this with your D&T teacher.
- Check with your other subject teachers if you think they can help.

6 Working with other people

There may be parts of your Capability Task that could benefit from a team approach – carrying out a survey, collecting reference materials, brainstorming ideas, for example. You will need to organize these carefully so that everybody's task is improved.

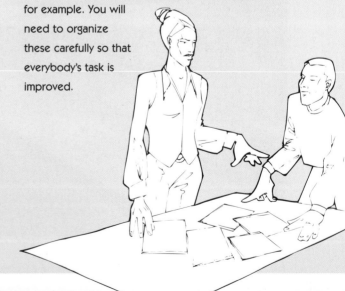

Tackling the Capability Task

7 Writing a design brief and developing a specification

You must remember that you are expected to design and make a quality product that meets demanding criteria. These should take into account how it could be manufactured, how it might be repaired or maintained and how it might be sold.

8 Generating design ideas

You will need to show where your ideas come from. Make sure you keep a record of your early thoughts.

9 Developing your ideas

You will need to keep a clear record of how your ideas have developed.

10 Making presentation drawings and working drawings

These should show what your design will look like and how it can be made.

11 Planning your making

12 Making your design

13 Evaluating the final product

Make sure you use a range of techniques.

14 Putting on a display

You should mount a display that shows your work to best advantage. It should describe the following:

- your ideas and where they came from;
- how they developed;
- presentation and working drawings;
- your schedule for making;
- your evaluation.

Writing your own Case Study

You may have to write your own Case Study as part of your GCSE assessment. Here are some guidelines.

Which product?

You should choose an everyday item that is manufactured. You should be able to examine it, use it yourself, see others use it and evaluate it. Here are some possibilities:

telephone   vacuum cleaner

refrigerator/freezer   radio

light bulb   hair-dryer

Walkman

What should it describe?

Your study should describe the following:

- what the product looks like;
- what the product does;
- how it works;
- who uses it and what they think of it;
- how it's manufactured;
- the product's impact on the way people live.

You might also describe:

- how the product has changed over time;
- other products that do a similar job.

How many words?

No more than about 2000 words. (One side of A4 paper filled with typing is about 500 words.)

What about pictures?

It is important to use illustrations as well as text. You can use any of the following:

- your own illustrations drawn directly onto the page or pasted in place;
- illustrations photocopied from books or magazines and pasted in place;
- your own illustrations scanned onto disc and printed in place;
- illustrations taken from a library on CD ROM and printed in place.

What about layout?

If possible use desk-top publishing (DTP) software to produce your Case Study. If this is not available use word processing (WP) software to lay out the text. If this is not available use a typewriter.

What about the overall length?

A reasonable mixture of text and pictures will give you a length of about 12 sides of A4.

What about special features?

You can make your Case Study:

- *attractive* by producing an illustrated cover;
- *easy to look through* by numbering the pages, using headings and producing a title page and contents page;
- *easy to understand* by using illustrations with notes and captions.

Examination questions

You may have to take a final written examination paper at the end of year 11 as part of your GCSE assessment. This paper will be made up of different sorts of questions. Here is a guide to these questions and how to answer them.

Interpreting a short Case Study

In this sort of question you will be given two or three paragraphs to read and one or two pictures to look at. The writing and the pictures will describe an aspect of design and technology from the real world. You will then have to answer a series of questions based mainly on what you have read. Some will involve finding a piece of information from the text. If you read the text carefully you can always get these questions right. Some will involve explaining something that is described in the text. These are more difficult as they will require you to use your design and technology knowledge and understanding. Some will ask you to make a judgement about the effects of the design and technology described. These are the most difficult but if you think carefully you will be able to use your design and technology awareness and insight to make judgements and give good reasons to back them up.

Presenting and interpreting information

In this sort of question you will be given data from some design and technology research and asked to present it in a way that makes it easy to understand. The data comes from very different sources. It could be about consumer preferences, the results of testing a material or component, production figures for different manufacturing methods or sales figures for different products. Once you have presented the data you will be asked questions that require you to interpret the data.

Why is it like that?

In this sort of question you will be given information about a product in the form of annotated illustrations and text. You will be asked to explain different features of the design such as:

- why a particular material has been chosen;
- why a part is the shape and form that it is;
- how particular parts work together to turn the input into the output;
- what would happen if certain things were changed;
- how particular parts might be manufactured;
- how the design might be improved.

You will need to use the design and technology knowledge and understanding you have gained in years 10 and 11 to give correct answers.

What could you use for that?

In this sort of question you will be given a short technical design problem. You will be presented with an incomplete design to which there are several different possible solutions. Your task will be in three parts:

1 to describe some of the possible solutions by means of simple annotated sketches;
2 to compare these solutions;
3 to state clearly which solution you think is the best with reasons.

Again you will have to use the design and technology knowledge and understanding you have gained in years 10 and 11 to give correct answers.

Questions 1 - 5 are taken from the RSA Examination Board GCSE Specimen Paper. They deal with design methods, material choice, mechanisms and manufacturing.

The diagram below shows an adult's Mountain Bike.

1 (a) **Name TWO** ways of finding out information that the designer of this mountain bike may have used when developing this product. *(2 marks)*

    (b) **Give TWO** features of this product that make it suitable for this age range. *(2 marks)*

    (c) **Choose TWO** different materials used on this product and justify clearly why they have been used. *(12 marks)*

Imagine that you are the designer who has been given the task of designing this product.

    (d) **What** is the purpose of the Design Brief for this product?
       (i)   Purpose of Design Brief
       (ii) **List TWO** items that must be included in the Design Brief. *(4 marks)*

    (e) **List SIX** points that would need to be contained in a detailed specification for this product. *(12 marks)*

The seat on this bicycle is made from rigid thermoplastic.

2 (a) **Name ONE** suitable thermoplastic for this purpose. *(3 marks)*

    (b) **Name ONE** suitable manufacturing process for producing this seat. *(3 marks)*

    (c) **Explain** how the chosen scale of production may influence the manufacturing process chosen. *(8 marks)*

Most bicycles manufactured today use mild steel for the frame.

3 (a) **Explain** why mild steel is the most commonly used material for producing cycle frames. *(2 marks)*

When deciding upon the most appropriate choice of materials for a design proposal the designer often has to resolve conflicting demands.

    (b) **Write** a short report explaining how a designer might justify using a material other than mild steel for a cycle frame. Your report should give **THREE** advantages and **THREE** disadvantages for a named material.

                                                          (7 marks)

There are a number of different mechanisms used in bicycles.
Figure 1 shows an incomplete front brake mechanism.

4 (a) (i) **Complete** this drawing by adding a suitable linkage to this system.

(*4 marks*)

(ii) **Add** arrows to your drawing to show which direction the separate parts move in when the brake is applied.

(*4 marks*)

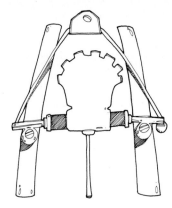

Figure 1

A company that currently uses repetitive batch production for producing its bicycle frames is considering investing in new computer controlled machinery to increase production.

5 Identify and **explain** the key considerations that need to be taken into account when making this decision.

(*11 marks*)

Questions 6 - 7 are taken from the ULEAC Examination Board GCSE Specimen Paper. They deal with choosing tools, batch production, material choice and design methods.

6 (a) **Complete** the table below by matching the fixing component with the correct tool. State the correct name for each tool.

(*8 marks*)

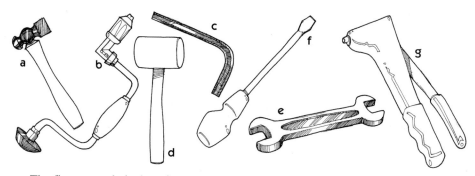

The first example is done for you.

Fixing component	Selected tool from above	Name of suitable tool
	f	Screwdriver

(b) **State TWO** safety precautions which need to be taken before switching on a drilling machine. It is set up to drill holes §6 mm in a flat plate of 9mm thick 100 square of milk steel. *(4 marks)*

(c) **Wood** is joined to wood using this type of screw.

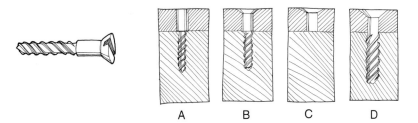

A   B   C   D

Complete the following statements. The first has been done for you as an example.

A   is incorrect because *there is no countersink for the head of the screw.*
B   is correct because

State a reason why C and D are incorrect. *(5 marks)*

(d) A right angled bend may be made from a piece of 3mm thick plastic using a strip heater. During heating, the plastic gave off a distinct smell and made a crackling sound. The surface of the plastic was filled with tiny holes and bubbles.
   (i)   State what caused this problem. *(2 marks)*
   (ii) State how it might be prevented. *(2 marks)*

(e) A piece of aluminium bar of 9mm diameter cracked and snapped when it was bent cold.
   (i)   State what caused this problem. *(2 marks)*
   (ii) State how it might be prevented. *(2 marks)*

7 The drawings below show THREE designs for desk tidies.

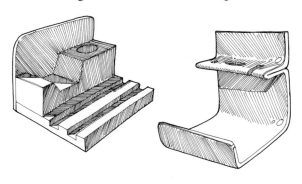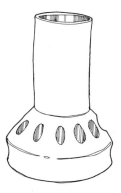

(a) All the desk tidies are made from different materials.
    State TWO suitable materials for making each desk tidy. *(6 marks)*

(b) **Choose ONE** design and **ONE** material and explain with reasons why the material you have chosen is suitable. *(3 marks)*

(c) **Describe** briefly how you would make the design you have chosen in part (b) *(6 marks)*

(d) **List THREE** important specification points to be considered when designing a desk tidy. *(3 marks)*

General Case Studies

Designing our surroundings

Our surroundings and the buildings we live and work in play an important role in how we feel about ourselves, and the world we live in. People work better if they have an environment which is comfortable and stress free.

Architects design environments to help people work better. When designing a new building the architect will consider factors that affect people such as:

- the air they breath;
- the opening and closing windows;
- the temperature;
- the lighting, both natural and artificial;
- sources and level of noise;
- the closeness of other people.

Pause for thought

- Try to remember an occasion when you felt uncomfortable in a room or building. What was it that made you feel that way?

Designed for work

When the architects were designing the Powergen building in Coventry they wanted to make it as energy efficient as possible as well as a good place to work. They decided to use a computer system to monitor the temperature, air flow and lighting. The computer uses the data it receives from sensors around the building to keep a constant check on all these things and to adjust them to save energy. It can open and close windows, turn on lights, etc. when the data it receives indicates that this is necessary. However, individual workers can override the computer, if they want to, at any time. This makes people feel happier because they are in control of their environment and the computer system is still able to save on wasted energy.

The architects were also asked to find ways to improve people's ability to work. Some modern buildings have been found to make people working there feel sick. **Sick Building Syndrome (SBS)** has been associated with factors such as air-conditioning systems, bad lighting and lack of building hygiene. Architects now know more about SBS and the subsequent rise in standards of materials handling during construction has largely eliminated SBS from new buildings today.

Questions

1 Make a list of the environmental factors that affect whether you can settle down to do your homework.

2 Make a list of the environmental factors that might affect whether an office worker can work efficiently.

3 Compare your two lists to identify the things they have in common.

The environment inside the Powergen Building is computer controlled

Designed to prevent crime and vandalism

When the Docklands Light Railway was being designed the architects knew that the railway went through tough, crime-ridden areas of London and that many of the stations were to be unmanned during large parts of the day. Any solutions they proposed would need to meet three targets:

- prevent crime;
- improve passenger safety;
- require minimum maintenance and repair.

Very little glass is used in the station designs as glass is often vandalized and needs replacing regularly. All the materials used in the construction of the stations have been chemically treated to make graffiti easy to clean off.

R

Research activity

Make a sketch (or take a photograph) of a building, and its surroundings, that is poorly designed or misused. Add notes to show what is wrong. List some of the ways which would improve the environment for everyone living or working there.

These designs have helped make Docklands safer

A 45 m lighting mast casts a brilliant light across the whole area and is designed to resist vandals climbing the mast or throwing missiles.

The gull-wing platform canopy is designed to be attractive and functional. It is made of tough poly-carbonate which is scratch-resistant, strong and tough. It lets light through but is hard to climb onto and difficult to damage.

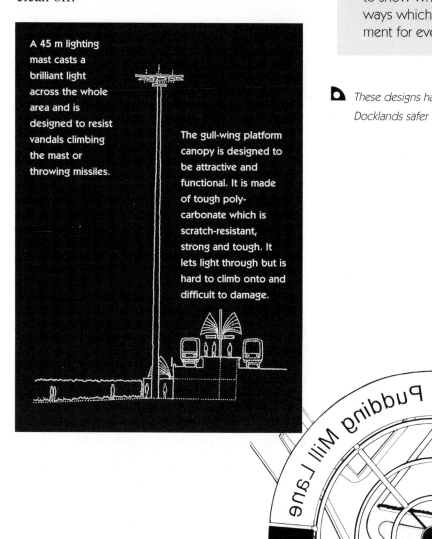

Looking down from the top of the lighting mast note the thorny hedge next to the path that leads to the ticket booth. This protects passengers from ambush by muggers. The ticket booth is designed to resist ram raiding even by a JCB.

Designed to entertain

London's Planetarium is Britain's largest and one of the capital's major tourist attractions. Its large green dome in Marylebone Road is a familiar London landmark. The Planetarium presents a dramatic exploration of the universe. It was built in 1956 but was not designed for the large numbers of visitors it now attracts – up to 4000 per day in peak season.

The project for upgrading the Planetarium for the twenty-first century was well over-due and it had to be completed in just 20 weeks to reduce the amount of time it was closed, and to minimize disturbance to Madam Tussaud's.

The architects came up with a practical solution which retained all the drama and expectation associated with such an attraction.

R **Research activity**

Take a trip to a local public attraction or entertainment venue. It can be a museum, cinema or theatre. Make a sketch or take some photographs and add notes to explain the following:

*   how the building is laid out;
*   how lighting and sound are used;
*   how people are moved around the building.

From your observations make a list of possible improvements.

Curved corridors entice the visitor into the auditorium. The dramatic effect is enhanced by sound and lighting. This also keeps people moving, which is important when dealing with a large volume of people.

◖ *The London Planetarium from the outside*

The original round theatre seating has been replaced with seats facing in one direction only. The seats are now raked to give people a better view, and have been designed by computer to calculate accurately the sightlines for each seat.

A new Digistar laser projection system has been installed to provide state-of-the-art images.

The projection dome has been raised to allow more space underneath for people to queue inside the building and for a new floor to be inserted. This will reduce congestion on the pavements outside.

◖ *The new design for the inside of the London Planetarium*

Information – the power to change lives

The way we communicate with other human beings and the speed with which we receive information have influenced dramatically the kind of world we live in today.

The start of books

Five hundred years ago most people were illiterate and relied upon word of mouth to receive information. Books were only available to closed religious orders and were laboriously copied by hand. In 1448 a German goldsmith, Johan Gutenberg, invented a way of printing whole pages using movable type. Gutenberg used this method to produce the first printed bible in 1456.

William Caxton was the first Englishman to develop a printing business. He printed his first book in the English language in 1474, called *The Recuyell of the Historie of Troye*.

Pause for thought

What is the possible connection between books becoming more widely available and schools opening?

As printing became faster, books were published on all sorts of subjects. As more books became available more people learned to read. It was also around this time that the first schools were opened.

How books shaped people's ideas

In France, the illegal and elicit distribution of cheaply printed books called 'Chap' books became commonplace by the late 1700s. One of these Chap books informed the ordinary people about the enormous excesses of the monarchy. Chap books helped to kindle a sense of national identity which ultimately lead to the French Revolution and the overthrow of the monarchy.

Pause for thought

Can you think of examples where the press is criticized for printing stories about royalty ?

▶ *Information from books added to the resentment that caused the French Revolution*

Information becomes important for business

In 1605, the year of the Gunpowder Plot, the first newspaper was printed, in Antwerp, Holland. The first British newspaper was called the *Daily Courant* and first appeared in 1702. At that time newspapers were read mainly by businessmen and merchants because they contained stories from other parts of the world which were important to trade and politics.

Questions

Working in a group discuss the following.

1 With new industry new jobs develop. What jobs do you think were created to operate the newspaper industry?

2 How do you think newspapers got their news two centuries ago as compared with today?

3 **a** Who decides what news is printed in the newspapers?

   b What might influence their decisions?

Pause for thought

Despite technological advances news is still passed on by word of mouth today. Journalists often interview on-the-spot witnesses or experts before writing up the stories for the newspapers.

New inventions and discoveries for mass communication

By the end of the nineteenth century Sir Alexander Graham Bell, a Scottish scientist, had invented the telephone and Guglielmo Marconi had invented the radio. It was now possible to communicate rapidly with people across the world. Radio had such potential for mass communication that the government set up the British Broadcasting Company in 1922, later to become the British Broadcasting Corporation (BBC).

Radio made it possible for people to hear the voices of politicians and other important people for the first time. The people of Britain would have heard Neville Chamberlain's famous announcement that Britain was 'now at war with Germany' in September 1939 on the radio.

By the mid 1930s nearly every family had a radio – for many a major source of news and entertainment. Radio stars were as popular as pop stars are today

Radio with pictures

During the 1920s John Logie Baird, another British scientist, was working on the idea of talking pictures. The BBC immediately became interested in his idea and started to develop television, transmitting its first programme in 1929. However it was not until the coronation of Queen Elizabeth II was televised in 1953 that large numbers of people hired TV sets and started to watch television regularly.

In the 1970s satellite communications were developed so pictures could be transmitted as they happened from anywhere in the world.

Pause for thought

How much time do you spend each day listening to the radio and watching television?

Questions

Working in a group discuss the following.

4 Do you think that television and radio have changed our lives for the better or the worse?

5 Do you think television influences people's opinions?

6 Do you think it right that the government can censor what we watch?

Mass communication has changed our lives forever

Television is thought to have a very strong influence on people because of its power to shape our thoughts and ideas. The government regulates what we see on television and has the right to veto a programme if it thinks the content is not in the national interest.

At first there was only one channel and pictures were transmitted in black and white. By the 1960s ITV, the first commercial channel, had been given a licence to broadcast and later, in the mid 1960s, colour was introduced

▶ Reporters covering wars, the World Cup or the Olympic Games all use satellite communications, so we can see events as they happen, wherever they are taking place in the world

ℝ Research activity

Find out about access to the Internet in your area by answering these questions.

1 Can you log onto the Internet at home?

2 Can you log onto the Internet at school?

3 Can you log onto the Internet in the local library?

4 Do your parents log onto the Internet where they work?

5 Is there an Internet users service for hire in your town?

Use the answers to these questions to comment on how much the Internet is used in your area.

Information is power

Today almost everyone needs information for work, leisure, education and for the day-to-day running of our lives. We can access the information we need very quickly, almost instantly in some cases, through using information technology. A single CD ROM can hold the information of many encyclopaedias, and high street banks' on-line computers can give instant information about personal finances at cash points throughout the country.

Having relevant information enables people to make decisions and to have more control over their lives. By using computers, phone lines and satellite links, the Internet allows people to exchange ideas and information with anyone in the world. It lets people communicate cheaply and rapidly without the information being edited or censored by a publisher, broadcaster or government. Communication via the Internet is a two-way process, meaning that anyone who transmits information on the Internet can have a dialogue with whoever receives the information anywhere in the world.

▶ Using a CD ROM to access information

DIY medical testing

Some products could not be designed if the designers didn't understand the science behind the way the product works. Obvious examples are motor cars, radios and televisions, microwave ovens and thermal blankets. The science behind these products is mainly physics and chemistry. Now new medical products are being developed which depend on an understanding of biology. For example, in the past a doctor would test for diabetes by sticking a finger into a sample of urine and licking it to see if it tasted sweet. Nowadays the doctor would use a chemical test strip developed specially to test for sugar.

 Old-style medical testing

Some medical tests are so simple and reliable that anyone can carry them out. A new type of product has therefore come onto the market – do-it-yourself medical testing kits.

P Pause for thought

What medical conditions might people want to test themselves for?

Testing for pregnancy

When a woman becomes pregnant she produces a chemical called human chorionic gonadotrophin (hCG). This chemical is present in a woman's urine when she is pregnant. In order to find out if she is pregnant, therefore, a woman can test for hCG in her urine.

Until fairly recently the only way to test for hCG was to inject the urine sample into a female animal such as a mouse or toad. If hCG was present then the woman's urine would cause the animal to produce eggs. If she was not pregnant there would be no hCG so no eggs would be produced. This test had many disadvantages:

- it had to be carried out by a laboratory technician;
- it took several days;
- sometimes the animals had to be killed to find out if eggs had been produced.

Biologists have discovered that our white blood cells produce antibodies as part of our defence system against attack by viruses, bacteria and certain chemicals, generally called antigens. The antibodies protect us by recognizing and combining with the antigens and rendering them harmless. The white blood cells produce particular antibodies to fight particular antigens.

In 1975 two scientists discovered how to produce large amounts of antibodies outside the body in a fermenter. This enabled scientists to produce a wide range of antibodies in large quantities, including one that could recognize and combine with hCG and nothing else. They knew this could form the basis of a reliable and accurate pregnancy test. Now it was up to product designers to develop an easy-to-use pregnancy testing kit.

Here's what they developed …

 The Clearblue One-step pregnancy testing kit

The pregnancy testing kit

To carry out the test a woman urinates onto the absorbent sampler.

There are two windows in the test kit. A blue line appears in the smaller of the two windows to show that the test is complete and has worked correctly. If the test is positive a line will also appear in the large window, showing that the user is pregnant.

How it works

How the test kit works is explained below.

R Research activity

Find out the meanings of the following terms:
monoclonal antibodies
hybridoma cells.

Q Questions

1 Why is the small window important?

2 Why is urine used for the test rather than blood?

3 a What are the advantages for a woman in knowing that she is pregnant as soon as possible after she has conceived?

   b Are there any disadvantages in knowing as soon as possible?

4 Why is it important for the test to be reliable and accurate?

5 Why is it important for the test to be easy to use?

6 What changes would you make to the test kit if it were to be used in a hospital laboratory?

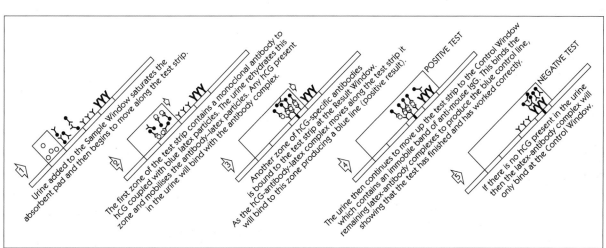

1 Urine added to the Sample Window saturates the absorbent pad and then begins to move along the test strip.

2 The first zone of the test strip contains a monoclonal antibody to hCG coupled with blue latex particles. The urine rehydrates this zone and mobilises the antibody–latex particles. Any hCG present in the urine will bind with the antibody complex.

3 Another zone of hCG-specific antibodies is bound to the test strip at the Result Window. As the hCG-antibody-latex complex moves along the test strip it will bind to this zone producing a blue line (positive result).

4 POSITIVE TEST
The urine then continues to move up the test strip to the Control Window which contains an immobile band of anti-mouse IgG. This binds the remaining latex-antibody complexes to produce the blue control line, showing that the test has finished and has worked correctly.

5 NEGATIVE TEST
If there is no hCG present in the urine then the latex-antibody complex will only bind at the Control Window.

3

Manufacturing aircraft

Since 1901, when the American Orville Wright made the first flight in a powered aircraft, the aerospace industry has come a long way. Early pioneers, sometimes working in garden sheds, built their flying machines out of linen and wood – during World War I furniture factories were enlisted to meet the demand for aircraft parts.

By World War II most aircraft were made of aluminium alloy, and it was car factories which turned to aircraft manufacture. Since then the development of passenger aeroplanes and increasingly sophisticated technology have transformed the aerospace industry. It is now vast, encompassing the design and manufacture of military and civil aircraft ranging from microlights and gliders to Concorde.

P

Pause for thought

Why do you think it was car factories not furniture factories that made aircraft in World War II?

Multinational manufacture

Every aeroplane is made up of thousands of different parts or components, all of which have to be functioning perfectly to ensure the efficiency and durability of the aircraft, and the safety and comfort of its passengers.

From the engine and wings to the door handles and headrests, every component has to be painstakingly designed, developed, tested and made. Because of the enormous amount of work involved, the parts to make one aeroplane are often produced in different factories all over the world before coming together for final assembly as shown below.

Designing and making the wing flaps

Shorts is one of Britain's oldest and largest aircraft manufacturers. It has been in the flight business since 1901, when Oswald and Eustace Short started making aerial balloons. Today Shorts employs 7800 people in the design, development and manufacture of a wide range of aircraft and aerostructures (aircraft components).

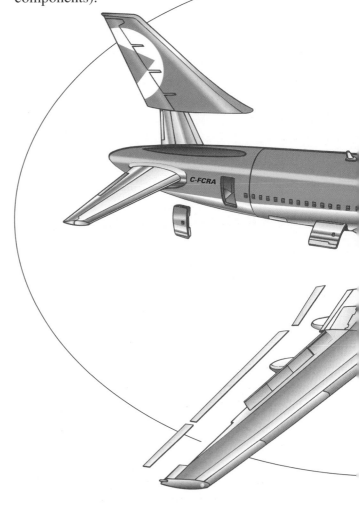

Shorts' aircraft and aerostructures are designed by Aero Designs Ltd on the Isle of Man. This is a special part of the company dedicated to designing aerostructures ready for manufacture and assembly at Shorts' sites. The designing is carried out using **computer-aided design (CAD)**. The results can be plotted out as 2D drawings, 3D wire frames or surface envelopes. From the design drawings the computer can calculate surface areas and volumes, and carry out stress analysis. All the members of the design team are networked to each other so that they can take into account how the others' designs are developing. For example, changes in the design of a wing will almost

certainly require changes in the design of the wing flaps.

Once the final design for the wing flaps is complete and the designers are satisfied that it meets the specification, the information needed to manufacture the parts is sent via a telephone line to the manufacturing site in Belfast. The information is fed directly into computer-controlled machines which can then be set to work to make the parts. This process of designing and manufacturing with the aid of computers is called **CAD/CAM** (computer-aided design/computer-aided manufacture). Once the parts are made they can be assembled into the completed wing flap which is sent to the USA for inclusion in the aeroplane.

Research activity

Find out if there are any manufacturing companies in your area. Make a list of them. Find out which ones use CAD to design their products, and which ones use CAM to make their products.

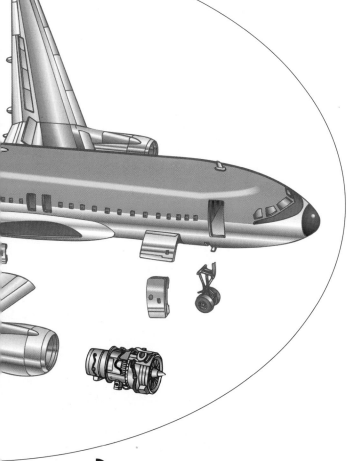

Parts for this airplane are made all over the world and assembled in the USA

CAD/CAM in action producing parts for a Boeing 757

Public transport In London

The early Victorians had rutted and cobbled streets with poor drainage and no road system as such. Today we have crowded buses, trains and tubes as well as an increasing number of vehicles creating more pollution and slowing traffic down. Are today's London residents and commuters any better off than their Victorian predecessors?

 Early victorian scene

Horses galore

The horse has played a key role in the development of public transport systems in cities across Europe. In 1829 the first regular horse bus service – carrying just 12 people – provided a fixed route service from the City of London to Paddington. It ran every three hours with a fare of one shilling (5p). This was a considerable sum of money at the time and so the service catered mainly for the wealthy. Its success encouraged other operators to set up services on other routes. This was the beginning of London's public transport system.

P **Pause for thought**

What is the future for public transport in London or any of Britain's other major cities?

Competition and cobbled streets

More operators in the market-place meant they had to compete for passengers. This made operators invest in ways of carrying more people. This resulted in the introduction of back-to-back seating on the tops of buses in 1850. These seats were accessed by a ladder – it took another 30 years to get a proper stairway to the top of the horse-drawn bus!

The size of the bus and therefore the number of passengers it could carry was limited by the power of the horses. A significant advance was made in 1861 with the introduction of horse-drawn trams. These had wheels which ran on steel tracks laid in the road. This made it easier for the horses to pull their loads which meant that buses could be made larger or more passengers could be carried on the existing buses.

 Travelling 1870

Pause for thought

What else might be used to power buses and trams? Why weren't these used in 1850?

Trams provided a cheaper form of transport with lower fares. This meant that more people could afford to use them to get to and from work. By 1900 many suburban areas of London were served by tram routes.

Powering the way forward

In 1906 there were around 50 000 horses working in London, transporting more than 2 000 000 people per day. But the days of horse-drawn transport were numbered. Towards the end of the nineteenth century operators explored other power sources, such as steam, electricity and diesel fuel. Their aim was to become more competitive by making their vehicles either faster or able to carry more passengers. Experiments with steam-powered trams were short lived and electricity proved to be the ideal power source. From 1901 overhead power lines or road-embedded conduit systems were installed.

The main problem with electric trams was that they were not very manoeuvrable. The power lines and tracks were often laid down the middle of the road. People had to dodge the traffic passing on either side to get on the tram!

Horses could not compete with these clean, quiet, reliable and larger capacity vehicles, and with the simultaneous introduction of the diesel motor bus, the horse's demise was complete.

The trolley bus: more manoeuvrable than the tram because it did not need rails

Trolley buses were introduced in London in the 1930s. They had already been used successfully in other cities for over two decades. They did not run on rails, though they were powered using overhead electric cables in a similar way to the trams. This meant they were more flexible as their manoeuvrability was limited only by the reach of their overhead power links. After World War II, from 1945 onwards the diesel-powered motor bus became the dominant form of public transport above ground and trolley buses ceased operation in London in 1962.

London electric tram in 1908

Going underground

London pioneered the underground railway system. It began with railway carriages pulled by a steam engine in 1863. This system was dirty and noisy and was replaced in 1890 by a deep-level electric system – the first in the world. Most of the central part of London's existing underground system was developed between the two wars, the outer reaches being developed after World War II. The London Underground is still being developed today with links to the Docklands Light Railway and the Jubilee line extension to the south of the River Thames.

Back above ground

The tram has not been totally forgotten. European cities such as Amsterdam have operated tram systems successfully for many decades and cities in England are beginning to follow suit. Manchester and Sheffield

Questions

1 As public transport became more comprehensive, reliable and affordable, London changed. People could consider living further away from their workplace, knowing that they could get to work by public transport.

What effect do you think this had on the areas surrounding London and in the Victorian slum sites within the city?

launched new tram systems in the early 1990s. London Transport is exploring the possibility of a Tramlink between Wimbledon and Croydon using existing rail lines. As a virtually emission-free form of transport and one that can maximize the use of rechargeable power sources, tram systems are an attractive proposition for congested and polluted cities.

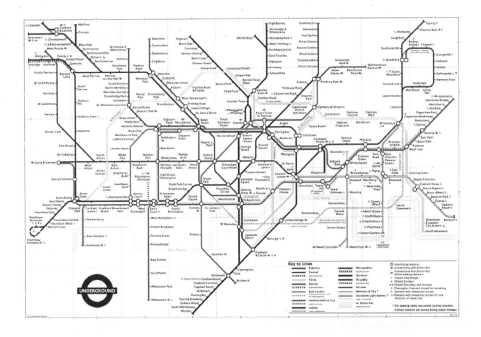

The modern tube map, first designed in 1933, shows only the sequence of stations and the connections

Ownership and intervention

Until the end of the nineteenth century individuals or companies owned various bus and tram routes. In 1891 London City Council started buying tramway companies and by 1899 they owned seven of the largest operations. By doing this the council was able to ensure that transport routes met the needs of the areas they served and the tram became the cornerstone of a public-owned transport system.

In 1933 the London Passenger Transport Board, a public body, was set up with powers to take over all bus, tram, trolley buses and underground services in London and adjacent counties. Today, bus routes are being offered for sale to private companies to manage. London Transport reported in 1994 'substantial savings of 15–20 per cent of previous operational costs … achieved from the tendering process.'

By the year 2001 it is planned that all bus routes will have been put out to tender to private companies. Over a period of 100 years public transport in London will have gone almost full circle, from private to public ownership and then back again.

Installing and managing the system

Public transport in London calls on all aspects of design and technology – civil engineering, mechanical and electrical engineering, large-scale manufacturing, advanced data capture and information handling systems, extensive maintenance and staff development.

London Transport has used the latest electronic and information technology to develop ticket vending machines and ticket reading machines linked to entry/exit points.

R

Research activity

Find out about the public transport in your area. Try to answer these questions:

1 What are the bus routes?

2 How frequently do the buses run?

3 What is the cost of a journey from the outskirts to the city centre?

4 What concessions are available?

You might present your information in the form of a display including an annotated map.

Modern designs keep improving transport in London

The 'look' of London Transport has been developed through corporate identity programmes that cover signage, uniforms, livery for vehicles, promotional materials and stationery.

3

Technological endeavours

Success or failure – pushing back the barriers

Throughout history scientists, designers and engineers have striven to push back the boundaries of invention and discovery. What is it that compels them? What are their efforts really worth? Often struggling against public opinion and a lack of resources, their achievements can have important effects on people's lives. Their failures often attract ridicule and scorn.

Not all research is carried out with a particular end result in mind – some research is purely academic. Designers and technologists can build on and adapt another person's discovery, creating a new product, system or material which was previously unimagined. It is often the long-term potential of design and technological endeavours, rather than the immediate results, that are of the greatest significance.

 Man walks on the moon – the culmination of years of endeavour

The space race

When American astronauts first walked on the surface of the moon in 1969, the event was heralded as a great achievement and the successful culmination of years of research, experimentation, testing and expense. But *why* did America want to put a man on the moon? Some see it as a struggle between the USA and the former USSR, each trying to prove that its political system produced the best technology. Once on the moon the astronauts took measurements and collected 400 kg of rock samples.

Questions

1 For each of the following highly successful products write down what the designer or inventor needed to know about to develop a successful design:

a Post-it
b PrittStick
c Tipp-Ex.

P

Pause for thought

It cost many millions of dollars for America to win the space race. Was it worth it?

The real value of the space race was, however, far more long term and influential. By landing a craft on the moon scientists discovered how to manoeuvre vehicles successfully in space. This led to further research which eventually produced satellites. Satellites now play a key role in people's everyday lives. Satellite-based telecommunications systems provide up-to-the-minute news, current affairs, entertainment and education via television pictures, radio and telephone links across the world. Navigation between countries and continents has been improved by the use of satellite information. Weather forecasts would not be nearly so accurate or informative without data capture by satellite.

Space race technology has allowed space stations to be placed in orbit around the Earth where research into solar energy, medicine and industrial processes are carried out. These experiments in turn could have unplanned and unpredicted far-reaching results.

Research activity

Find out what other 'spin-offs' there have been from the space race. Make a list and give each one a star rating for how useful it is.

One good thing leads to another

Scientists do not always reap the full benefits of their endeavours; once the purpose of their research is achieved, the project comes to an end. The further exploration and exploitation of the potential of each new development is often taken up by designers and technologists operating in other fields.

Teflon

One material used in the development of spacecraft provides a good example of how designers and engineers can pick up and adapt other scientists' discoveries.

In 1938 an American scientist working on refrigerants for DuPont, Dr Roy Plunkett, discovered a new material which he called Teflon, or polytetrafluoroethylene (PTFE). It had many unique qualities:

● resistance to high temperatures;

● very low conductivity;

● a very slippery surface;

● non-reaction with other chemicals.

However its commercial potential was not fully explored until 1954. Then a Frenchman, Marc Gregoire, used some PTFE to lubricate his fishing tackle and realized that the material could be spread over the base of a frying pan to create a non-stick surface.

The Tefal non-stick cookware company grew out of this discovery. Teflon later became an integral and important component of the spacecrafts sent to the moon – its resistance to high temperatures was essential for successful re-entry into the Earth's atmosphere.

In the 1990s Teflon is used by textile designers as a protective coating for garments. By applying a layer of Teflon to clothes it ensures that they are waterproof, breathable and stain repellent.

Pause for thought

What could the next application of Teflon be?

Microwaves

The development of microwaves also demonstrates how one achievement can be adapted and applied by other people. In 1940 two British physicists, Sir John Randall and Dr H A Boot, developed the magnetron, an electronic tube which produced microwave energy, for use in radar installations. It then contributed successfully towards Britain's war defences. Microwaves are radio waves less than 30 cm in length. They can be focused easily into a sharp beam which reduces the chances of interference from other transmitters.

It was the Americans and Japanese who explored the wider potential of microwaves in the domestic market-place. At one radar installation an American manufacturer called Percy Le Baron Spencer noticed the heat given off by a magnetron's electronic tube. He tested the strength of this heat by putting a paper bag full of maize into the field of the tube. Within seconds the maize swelled and burst, and Percy had perfect popcorn!

Spencer's firm, Raytheon, realized the commercial potential of the magnetron and set about developing a cooker operated by an electronic tube. Microwaves penetrate food to a depth of 50 mm and cause the water molecules in the food to vibrate rapidly. This makes the food hot.

The first Raytheon microwave ovens were large and heavy units meant for big users like hospitals and canteens. The technology continued to be developed and refined and the first household ovens were produced by the Japanese Tappan company. Microwave ovens are now one of the most highly desirable and widely available consumer items.

Questions

2 Make a list of some of the products you use every day. For each one write down the technologies that it uses.

The value of failures

Many technological endeavours result in failure. Not every design development can be a success, whether instant or long term, but the research and ideas that led up to the 'failure' can still be used and adapted.

Sometimes developments do not succeed because of the influence of direct competition. VHS and Betacam became available to consumers at the same time, and only one, VHS, survived in the market-place.

Sometimes designs are not taken up because the public is simply not ready for the idea.

Ultimately some ideas fail because they simply do not work. This does not make the endeavour worthless – other designers can learn from the mistakes and avoid them.

Research activity

Some of the most exciting technological developments in the future will depend on advances in our understanding of biology, particularly genetics. Find out about genetic engineering and suggest some products and services that may be based on this.

Focused Case Studies

Designing and making test equipment

Gearboxes in tractors

The gearbox fitted to a farm tractor is designed to be very strong and reliable. It is an important part of the transmission system which takes power from the engine to the driving wheels.

Tractors are expected to have much longer working lives than cars or vans. If a tractor breaks down, work is delayed bringing extra cost and inconvenience to the tractor's owner.

P

Pause for thought

Why might tractors be expected to last longer than cars?

A newly built tractor gearbox must leave the factory in good working condition. Any faults, such as badly aligned gears which will not mesh properly, must be identified by the manufacturer and corrected. This reduces the chances of the new tractor breaking down.

Investigating the nature of any faults enables a manufacturer to take 'corrective action' to improve the overall quality of the product. The detection of a faulty gearbox is very important and justifies the design and construction of a special test rig at considerable cost to the manufacturer.

P

Pause for thought

What other parts of the tractor might be important enough to justify special testing?

◖ *Gearbox mounted in rig*

How to test the gearbox

Experience in manufacture has shown that a defective gearbox will not transmit power efficiently. The test equipment specifiers, Irwin Desman Ltd, therefore decided upon a **dynamometer** test rig. Dynamometers measure power. The rig would need to drive the input shaft of the gearbox with a powerful electric motor to simulate the tractor's engine. At the same time it would apply a load to the output shaft with an electrical brake. (This is similar to an electric motor which is driven to oppose the rotation of the output shaft.) The speed of rotation of the output shaft and the torque (turning force) applied to the brake by the output shaft would be measured by instruments fitted to the rig. These readings would give an indication of the power (work done per second) being delivered by the gearbox. In simple terms if the brake could slow down or stop the output shaft easily the gearbox was defective.

b *Instrument dials on test rig*

Questions

1 Investigate the power output of a small electric motor by finding out how much load is required to just stop it turning.

Designing the test rig

The test rig had to be designed so that gearboxes could be fitted to it and removed easily, with no obstructions which might slow the operation down. The test results had to be displayed clearly, using instruments which were sturdy enough to operate in an oily working environment, possibly with high levels of electrical interference and mechanical vibration.

The safety of the test rig operator was an important consideration. To ensure the operator was not injured by coming into contact with moving parts, the controls were designed so that both hands had to be used for operation. Sheet metal guards were designed to protect onlookers from moving parts. (These guards have been removed in some of the photographs, for clarity.)

A full set of engineering drawings was produced from the initial design layout, and a full list compiled of all the standard components such as the electric motor, control switches and the dynamometer itself. Once this was done, a prototype rig could be constructed and tried out.

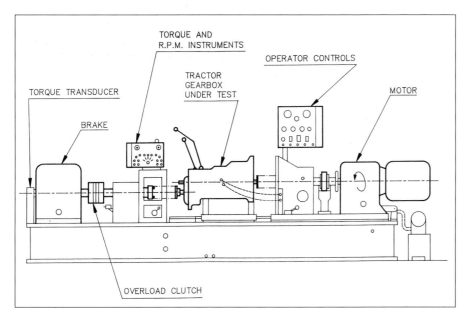

Testing the test rig

Initial trials showed the need for some minor changes to the design of the test rig. Extra sheet metal trays were added to catch the oil which spilled from the gearbox when it was removed after testing. A sound level meter was used to measure the noise levels around the prototype test rig whilst it was running. This indicated that sound levels had to be reduced so as not to cause discomfort to the operator. Extra panels were therefore fitted around the motor to act as acoustic screens. When new tests gave satisfactory results, the original drawings were brought up to date to incorporate the successful modifications. The rig was tested and proved in the designer's factory before being shipped out to a factory in Antwerp where the gearboxes are made.

▷ *Schematic diagram of test rig*

The test rig in use

After extensive trials in the gearbox factory, the test rig was judged a success, but the time taken for testing created a 'bottleneck' in the production process.

To solve this new problem, two additional test rigs were ordered. These could be built and delivered easily, making use of the full set of proven drawings which the designers had produced.

R

Research activity

Find out how the following items of equipment are tested:

car brakes
car safety-belts
electric motors used in tape and
CD players.

▷ *Overall view of test rig*

45

Baby Buggy

A transferable idea

What has the undercarriage of the classic World War II aeroplane the Spitfire got to do with baby buggies? Owen Finlay Maclaren designed both! He applied his knowledge of complex 3D folding structures, used in the development of the Spitfire's undercarriage, to the problem of transporting his three infant grandchildren.

Maclaren's finished design provided a lightweight fold-up child transporter that could fit in the boot of a car or in a cupboard when no longer needed. To achieve this his design solved a very difficult three-dimensional fold problem which today's designers would have modelled using CAD systems. Maclaren developed his idea on paper using sound engineering principles and experience gained from his Spitfire design days. His umbrella-fold design, as it has become known, was launched in 1965, and it has become a design classic.

The Maclaren Baby Buggy marked a dramatic departure from the large heavy prams and pushchairs available at the time and has had a profound effect upon the way in which pushchairs have developed.

To comply with British Standard requirements, all models have a minimum three-point harness built in. More expensive models have a five-point harness which offers greater security by providing additional shoulder straps.

The basic buggy weighs just 3 kg. Even those at the top of the range weigh just under 7 kg. Double buggies weigh in at 10.4 kg.

The buggy now features a pair of wheels at each corner for extra strength and stability, straighter tracking and a smoother ride. Depending on the model, some have fixed wheels and others have swivel wheels for increased manoeuverability.

The seat position determines the age of child that can be carried in the buggy. Babies under six months have to be able to have a lie-back position beacuse they have limited neck and body strength. The basic buggy has a fixed seat position which means that it can only be used with children over six months old. Some Maclaren buggies offer two adjustable positions, including lie-back, and others provide up to five different positions for extra flexibility.

P

Pause for thought

How many different models of umbrella-fold buggies does Maclaren manufacture? How many buggies do you think Disneyland Paris ordered? (The answers are 10 and 1700 respectively.)

R

Research activity

Collect information about a Silver Cross pram or a large pushchair from the 1960s and compare this with the Maclaren Baby Buggy. Make a list of the major design differences. Explain how two such different products can be marketed for the same main purpose – transporting babies.

The buggy comes in a range of different fabric pattern options, from very colourful and bold to plainer, more restrained designs. They are hardwearing, stain-resistant and easily wiped clean.

Optional extras – for increased flexibility the buggy has extras which can be purchased separately; these include a foot muff for really cold days, a rain hood and a sun canopy. So there is something to protect the baby from all sorts of weather.

Brakes are an essential safety feature and have to be applied when parking the buggy, when a child is getting in or out, or when the buggy is being folded up or opened. They are located so they can easily be applied by your foot.

A commercial success

The Baby Buggy is now a household name – people often refer to this type of pushchair as a 'buggy' whether it's a Maclaren or not! The buggy provided the foundation for a company which now manufactures 500 000 buggies a year, 25 per cent for export, plus eight other ranges of prams and pushchairs which are constantly being updated. All of these contribute to the company's 30 per cent share of the UK baby transport market.

R

Research activity

Find out about the Duette and the Pixie – two products that have been developed from the Maclaren Baby Buggy. Draw up a table showing how they are similar to and different from the basic buggy. Use this information to work out who is likely to buy each model.

 The buggy folds in three dimensions

Q

Questions

1 **a** Use the sequence of pictures to understand how the Maclaren Baby Buggy folds up.

   b Make a series of drawings or a simple working model that explains how it works. Remember that when it is folded up the structure is thin but as you open it, it gets wider. The structure is a shape-changing mechanism!

3

The jewellery business

This gold jewellery was excavated from a site in Cyprus in 1896 and is over 3000 years old. This style is still evident in jewellery today.

Jewellery has existed in many forms over thousands of years. These have included brooches, clips, hair-slides, rings, earrings, tiaras, necklaces, girdles and bracelets, some of which would often have amulets or pendants hanging from them. Early cultures often used precious metals and gemstones in their jewellery.

I am what I wear?

Jewellery today is used on many levels to express different things. For some people it is a statement of wealth – I am wearing diamonds and platinum therefore I am wealthy. For others it can reflect their interest in fashion – they wear jewellery that is contemporary and in the style of a particular fashion trend. For others it might be about wearing an item that is unique, perhaps made by a designer-maker; something that is individual and that you will not see other people wearing. Or it might be simply that you just like something or were given it as a gift. Usually the choice of a piece of jewellery will be for a mix of these reasons.

Prices for jewellery vary enormously, from earrings for under £1, to jewellery made with precious metals and gemstones costing many thousands of pounds. In the high street, jewellery can be found in most fashion stores and traditional jewellers. This is the type of jewellery most of us can afford to buy.

Jewellery takes many forms and uses a very wide range of material – solid silver and gold; less expensive metals plated with silver and gold or covered with plastic; anodized aluminium and titanium; inexpensive metals such as copper and brass; plastics and even leather and paper.

Major consumer markets

The world market for jewellery is estimated to be in the region of $US120 billion. This is an estimated total of consumer spending on all types of jewellery – so the items you buy are represented in these figures. Of the total of $US120 billion, just over 67 per cent ($US80.8 billion) of this market is shared between seven leading players.

The jewellery market

Country	Number of items sold	Value ($US)
USA	92 million units	34 billion
Japan	68 million units	17 billion
Italy	63 million units	11 billion
Germany	43 million units	5.5 billion
United Kingdom	83 million units	4.8 billion
France	49 million units	4.3 billion
Spain	44 million units	4.2 billion

Figures reproduced with kind permission from Grant Walker Ltd.

Knowing the market

As with any product, whether domestic appliances, cars or jewellery, knowing the market is essential. This means knowing:

- what sells;
- what is available;
- how much it costs;
- what are the emerging trends in materials, style or colour.

The jewellery industry holds many trade fairs all over the world. These provide a range of opportunities for those who work in the jewellery industry:

- promotion of new ranges to buyers and potential outlets;
- the latest technical information about new materials for jewellery;
- up-to-date information about new techniques for jewellery manufacture;
- the latest information about the cost and sources of materials and components;
- showcases for innovative designers.

Trade fairs provide an overview of what is going on in the industry and help to keep people in the trade informed of the latest developments. The main trade fairs such as Vincenzaoro (Vicenza, Italy), International Jewellery (Tokyo, Japan), Inhorghenta (Munich, Germany) or JA International Jewellery Show (New York, USA) are held in the key market countries.

Questions

1 a Using the figures in the table above calculate the average unit price spent, in $US, for each country.

b Present this information graphically.

c From this information what conclusions can you reach about the jewellery market in the UK compared with other countries?

Designing jewellery worldwide

Grant Walker provides design consultancy to the fine jewellery industry. Its companion company, Walker Walker Grant, works within the fashion and costume jewellery industry. Both companies offer design, manufacturing and marketing consultancy, as well as forecasting and trend prediction.

Original designs are created for clients, both large and small, all over the world. This may require the development of an entire collection or just one or two examples for presentation or marketing purposes. Clients range from Next and Oasis, providing low-cost quality fashion and costume jewellery for the high street, to manufacturers of fine jewellery using precious metals and gemstones.

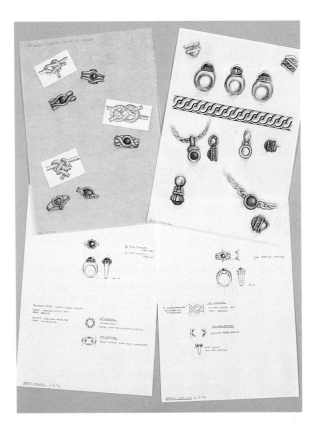

The drawings take knots as a starting point and develop this into a range of ring designs (Courtesy of Grant Walker)

Setting up a large-scale jewellery business requires information and judgement

R

Research activity

Visit some local shops and collect information that you think will enable you to predict what jewellery will be the most popular in a year's time. Present your findings as a short report.

In both fine and fashion jewellery, staff have particular areas of expertise in casting, stamping, electroforming, chain-making, polishing and finishing, electroplating, alloying, assembly systems, master-making, stone-setting, precious metals and CAD/CAM applications. Recently, they developed a complete feasibility study for establishing a new manufacturing unit in India. This involved researching factory sites; specifying the size of building and layout of internal spaces, the equipment required, staffing levels, training programmes and manuals for staff; advising on materials; forecasting potential market share; market intelligence on competitors; branding advice and more. So a new entrant to the market can have a complete picture of the capital investment required, the risk attached and an idea of the possible growth and potential of the chosen market area.

Grant Walker also works on projects with aid agencies which aim to support developing countries, for example India, in developing their own economy by maximizing their already vast but mainly domestic jewellery market. This requires training programmes for employees and advice for companies about equipment and the importance of design when competing in world markets.

Two designer-makers

These two designers work in quite different ways using very different materials but they have a common bond – they are both designer-makers. They produce jewellery in limited runs and also some one-off pieces, often to a commission from a private client.

Anne Finlay

Anne explores geometric shapes to achieve eye-catching and innovative forms. She usually models her ideas in 3D straightaway rather than sketching them on paper. She works primarily with plastics using PVC, nylon and acrylic. This enables her to create lightweight, colourful jewellery.

Thick pieces are cut out using an engraving machine. They can be assembled into even thicker pieces by laminating. Thin, flexible PVC shapes, some of which have been screen printed with geometric patterns, are either die-cut or hand-cut using a scalpel and templates. They are then combined with other materials such as stainless steel and rubber. Functional parts such as brooch pins or earring posts are incorporated as integral parts of the design.

Anne produces runs of between 10 and 25 at a time. Her pieces are sold to shops and galleries in the UK, USA, Europe and Japan, through exhibitions and also by mail order.

Jane's 'Byzantium' range of jewellery includes clocks, coasters and mirrors, all decorated using monoprinting

Jane Adams

Jane's work exploits the potential of aluminium. She has evolved her own techniques of painting, printing and dyeing the surface of anodized aluminium to create intensely coloured and patterned pieces. The shapes and forms are based on natural ones: shells, fish or birds, and the highly decorated surfaces are inspired by patterns and textures from carpets and embroideries.

For her production runs, Jane usually prints the pattern onto the aluminium using special coloured inks and rubber stamps. Sometimes she screen prints her surface pattern directly onto the aluminium. For her one-off pieces Jane hand-paints or draws on the desired pattern and in some cases finishes the piece with gold leaf.

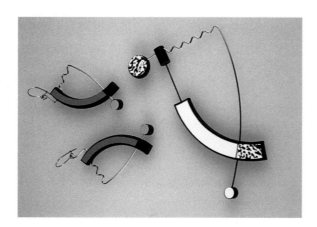

Colourful, lightweight matching brooch and earrings

Making light work

Today we take good lighting for granted, expecting it not only to be efficient and cheap, but also to meet our needs and moods. Yet until about 150 years ago, the only ways you could light a room were with candles made of beeswax or tallow (animal fat), or with lamps burning animal or vegetable oils.

P

Pause for thought

What would your life be like without electric light?

As the timeline below shows, fuel has been the key to advances in lighting technology. The discovery of paraffin in 1850, followed by the harnessing of gas as a lighting fuel, marked major breakthroughs. But it was not until this century that electric lighting transformed our everyday lives. Even then, many rural areas of Britain still relied on gas and oil lighting as recently as 25 years ago.

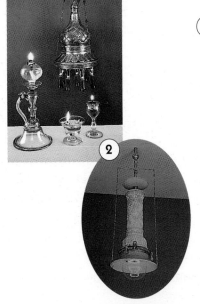

Lighting through the ages	
50 000 BC	First oil lamps, made from stone and using animal fats as fuel.
3000 BC	Candles were in use in Egypt and Crete.
Early 1500s	First attempts at street lighting: candle lanterns hung outside houses.
1681	Oil lamps first used for street lighting.
1784	Oil lamp designed with a hollow wick, making the oil burn brighter.
Mid to late 1800s	Use of paraffin and gas as fuels for lighting grew.
1879	Thomas Edison produced the first electric light bulb.
1909	Tungsten filament used instead of carbon in light bulbs.
1910	First neon light produced.
1935	Fluorescent lights first demonstrated.
1945	First fluorescent light installed in Britain at Piccadilly tube station.
1959	Mains-voltage tungsten halogen lamp developed.
Early 1980s	Low-voltage tungsten halogen lamp developed.
Early 1990s	Mini fluorescent lamps developed, suitable for use in the home.

The lights around us

The **incandescent light bulb** was the first to be designed, and is still the most popular today. When electricity passes through the tungsten filament, the wire glows brightly at a temperature of more than 2500 °C. Inert gas inside the bulb prevents the wire from oxidizing. Domestic light bulbs are designed to last for 1000 hours. They produce a warm, natural light but aren't very efficient, emitting a lot of heat as well as light.

Fluorescent light is produced by passing electricity through a phosphor-coated tube containing mercury vapour. An electrode at the end of the tube emits electrons. These collide with the mercury atoms causing them to emit invisible ultra-violet light. This light makes the phosphor coating produce a fluorescent glow of visible light. Fluorescent lights are longer lasting and more energy efficient than standard light bulbs. They do not get hot but tend to produce a colder, less natural light.

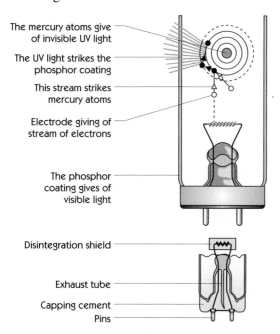

The mercury atoms give of invisible UV light

The UV light strikes the phosphor coating

This stream strikes mercury atoms

Electrode giving of stream of electrons

The phosphor coating gives of visible light

Disintegration shield

Exhaust tube

Capping cement

Pins

Fluorescent light tube

Ⓡ Research activity

1 Find out what kinds of light sources are used for street lighting, lighthouses and to light film sets.

2 Investigate neon lighting and draw a diagram to show how a neon light works. How can you vary the colour of a neon light? In the late 1930s there were hundreds of firms producing neon signs. What alternative has replaced it today?

3 Domestic light bulbs in isolated houses may last twice as long as those close to electricity substations, where voltage is up to 6 per cent higher. Find an explanation for this.

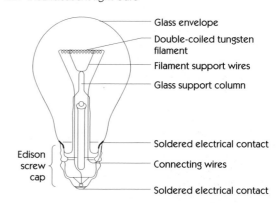

Incandescent light bulb

Glass envelope

Double-coiled tungsten filament

Filament support wires

Glass support column

Soldered electrical contact

Connecting wires

Soldered electrical contact

Edison screw cap

Mains-voltage tungsten halogen bulbs last twice as long as conventional bulbs. Evaporated matter from the tungsten filament reacts with halogen gas and redeposits itself on the filament rather than blackening the bulb. A **low-voltage tungsten halogen bulb** was developed in the 1980s, which runs off 12 to 24 volts (as opposed to the mains voltage of 240 volts) and is small enough to be recessed into a ceiling. A transformer is needed to convert the electric current from mains to low voltage.

Lighting design

Lighting designers have two main concerns:

- providing the right type and brightness of light;
- designing the light fittings.

Getting the light right

Lighting plays an essential part in our day-to-day lives, and thanks to the invention of the electric light bulb we now have instant light at the flick of a switch. Artificial light makes the day longer and gives us comfortable, efficient vision. It defines texture and colour, creates atmosphere and helps us to work and relax.

It even affects our moods – scientific research has shown that lack of bright light can lead to depression.

As technology develops, more and more options are open to lighting designers. Their priority is to design lighting systems which meet the demands of a particular environment – from a clear, practical light for working, to a soft, natural light for relaxing. Getting the balance right involves research into how people use particular spaces. For example, a library combines high levels of overhead illumination so that people can find books easily, with concentrated lighting at reading desks so people can focus on specific tasks. In this case, the lighting designer might specify overhead fluorescent tubes with standard light bulbs in desk lamps.

In public spaces, lighting is often used to try to influence people to behave in a certain way. Fast-food restaurants tend to be brightly lit to encourage movement, whereas smaller restaurants often use subtle, shadowy lighting to encourage customers to linger.

Different situations require different types of lighting and styles of fitting

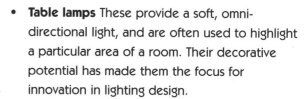

To help sell goods and to deter shoplifters, supermarkets have a high level of overhead illumination, often accompanied by smaller light sources near particular foods to bring out the colours of the packaging.

Fitting designs

Ever since people first used candles and oil lamps in the home, light fittings have been designed for both practical and decorative reasons – from crystal chandeliers to simple ceramic candleholders. However, in the past lighting designers were restricted by the limitations of candles, oil and gas. It wasn't until the introduction of the electric light bulb that designers were free to experiment with a wide range of materials, shapes and styling.

Light fittings fall into four broad categories.

- **Ceiling lighting** The most common forms of ceiling lamps are hanging pendants, ceiling-mounted globes and downlighters, which are recessed into the ceiling to direct light downwards. All provide general, omni-directional light.

- **Floor and wall lamps** These include standard lamps, which emit an omni-directional glow; wall-washers (usually ceiling mounted) which beam light evenly across a wall; and wall-mounted light fittings which diffuse light into a room. Both wall and floor lamps are often uplighters – light fittings designed to beam light upwards where it is reflected from the ceiling.

- **Table lamps** These provide a soft, omni-directional light, and are often used to highlight a particular area of a room. Their decorative potential has made them the focus for innovation in lighting design.

- **Desk lamps** These are specifically designed to light a particular task or work area. The Anglepoise light, the best known and first adjustable desk lamp, was launched in 1934.

Most rooms use a combination of these light fittings to suit the range of activities people carry out there at different times of the day.

Questions

1 **a** Draw a plan of your ideal bedroom, including a sleeping area, a working area with a desk, and an area where you can relax, watch TV or listen to music.

b Now imagine you are a lighting designer and mark clearly on your plan where you would position lights.

c What types of light sources would you use (for example, standard light bulbs, coloured light bulbs, fluorescent lighting)?

d What types of light fitting would you specify?

Research activity

Survey the lighting around your school. List the different types of sources and fittings used in each area and evaluate how effective they are in terms of aiding vision and aesthetics.

More than just playthings?

For thousands of years, human beings have made and played with toys. Although novelties and fashions come and go, many toys have a long history. To most people nowadays toys are just playthings, designed to entertain and stimulate children. However, taking a closer look at three types of moving toys shows us that they have made a much larger contribution to the world we live in

P ## Pause for thought

What did you enjoy about the toys you played with as a child?

today.

Spinning tops and gyroscopes

Spinning tops have been played with by children all over the world for centuries, and they hold the secret to an important engineering breakthrough. If you balance a spinning top on the end of a pencil and then move the pencil, the top continues to point in the same direction. This is because the spinning top has **angular momentum** which

A collection of spinning tops you can buy today

keeps it pointing in the same direction unless a force acts on it.

Interest in spinning tops led directly to the development of the gyroscope. A gyroscope consists of a wheel-shaped rotor on an axle that is mounted inside a metal ring. The rotor is set spinning by winding a string around the

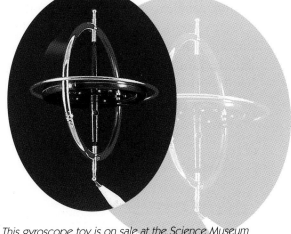

This gyroscope toy is on sale at the Science Museum in London

axle and pulling. Once spinning the gyroscope seems to defy gravity, resisting attempts to change its position.

The gyroscope remained a scientific toy until 1878, when an electric motor was added to keep it spinning. It was then that its practical uses emerged.

The most important of these were the gyrocompass and the gyrostabilizer. Today the navigational instruments and autoguidance systems of most aeroplanes and ships depend on some form of gyroscope. The gyrocompass consists of a gyroscope connected to an indicator with the axis set in a north–south direction. When the ship or aircraft turns, the

Research activity

Talk to your science teacher about the forces on spinning objects. Show him or her the pictures of spinning tops and ask him/her to explain why the mushroom shaped one turns upright as it spins. Use the explanation to draw an annotated diagram for a book about spinning tops.

model aeroplane, a glider derived from kites. Later toy aeroplanes and helicopters combined a kite-like wing and small tailplane with a propulsive airscrew behind. These were the direct forerunners of the first powered aeroplanes.

Optical toys

Nowadays we take it for granted that we can watch pictures move on film or television. But it wasn't until this century that people first understood moving images – and it would probably have been a lot later without Victorian optical toys.

The **thaumatrope** is based on the eye's ability to make two images appear to be in the same place at the same time. A disc with a different image on each side is rotated on thread, and the images are superimposed.

The **phenakistoscope** was the first optical toy to show animation. Images are drawn on a disc with slits around it. This is rotated in front of a mirror, and when you look through the slits the images appear to move. Cinema evolved from the principles of the phenakistoscope.

The **zoetrope** is similar to the phenakistoscope,

Questions

1 Based on the text and the picture shown below, design and make your own thaumatrope and flick books.

Research activity

1 Two main principles lie behind our ability to see moving images: persistence of vision and phi-phenomenon. Find out more about these, and design simple tests to demonstrate each.

2 The magic lantern, an early form of image projector using slides, was a forerunner of Victorian optical toys. Find out more about magic lanterns, and draw a sketch showing how they worked.

but was more popular because more than one person could watch the moving images at once. A paper strip showing sequential images is put inside a metal drum with slits around it. When you look through the slits as

Victorian and fun!

gyroscope keeps the indicator pointing north. In aircraft, the artificial horizon, which shows whether the plane is flying level or tilted to one side, is controlled by a gyroscope.

Kites, airscrews and flight

The kite and the airscrew, two ancient toys still popular all over the world, played a major part in the development of the aeroplane.

The kite has been put to many practical uses during its long history. In 1752 the American Benjamin Franklin proved the existence of static electricity in clouds by flying a kite with a metal key attached during a thunderstorm. By doing this he made a flash of lightning travel down the string to the ground. During World War I giant kites were used to lift men from warships to spot enemy ships over the horizon. (The chief kite instructor to the

British War Office was also the first man in Britain to build and fly an aeroplane.) The flexible kite, invented in 1948, is used in the US space programme to bring capsules back to Earth.

The airscrew used to be made of feather, wood or bamboo but is now usually plastic. It is launched vertically from a handle by pulling a string wound round its shaft or by spinning the shaft between the hands. In the early 1800s George Cayley experimented with airscrews and kites and established the three basic principles of flight – lift, thrust and control. He went on to design and fly the first

R Research activity

Find out about the different forms of kite:

eddy   box   tetrahedral
delta   parafoil.

Collect pictures of each and add notes explaining their structure and how they are controlled.

Still popular after many centuries

the drum rotates, the images appear to move.

The flick book, also known as the pocket cinematograph, is a series of sequential drawings bound as a book. When you flick through the pages, the images appear to move.

Modern manufacture

To be commercially successful across the world a toy will need to sell in the millions. So toys are manufactured using modern production technology. The mechanical animal shown below is made up of over 50 different plastic parts plus a clockwork motor. The plastic parts are produced by **injection moulding**. Molten plastic is forced into a metal block containing cavities that are the shapes of the parts. When the plastic has set

◗ *Assemble your own walking beast!*

R

Research activity

Make a collection of inexpensive toys that have been mass produced. Take them apart and make up a display showing how they have been manufactured.

P

Pause for thought

Here is the best-selling toy from Christmas 1994. What messages does it send to children?

solid the parts are released from the block. Using this method it is possible to make lots of different shaped parts very accurately and very quickly. The zoid toy is sold as a kit of parts which can be fitted together without the use of adhesives. The result is a mechanical animal which walks!

Getting the message

Although the main purpose of toys is to amuse and entertain, they can also be a powerful educational influence on children. For centuries society has recognized the potential of communicating knowledge, values and beliefs to children through the toys they play

Improving hygiene

In Africa, a readily available supply of water, plumbing and sanitation are confined to its major cities. In refugee camps and other crowded rural and outer-urban areas water is scarce and plumbing is usually non-existent. Toilets consist of a spot of open land or latrines, which are basically holes in the ground. The obvious health risks attached to this can be alleviated by providing more latrines and then using slabs to cover them over. These slabs may not be the most exciting product in the world but they can vastly improve the quality of people's lives.

A new design for an old problem

The latrine cover was pioneered in the early 1980s in Mozambique and western Tanzania. It incorporates a cast-concrete plug – reducing the smell and the flies – plus raised foot-stands and an inwardly sloping squatting area around the plughole, making it easier to use and keep clean. Obviously the social benefits of these features are enormous!

The slab is made in the shape of a dome rather than a flat plate so that it is strong enough without added reinforcing wire or rods. It can be made using simple manufacturing methods. The result is that the slabs are 20 to 40 per cent cheaper to produce than the flat slabs which used to be used.

P

Pause for thought

Why is a dome stronger than a flat plate of the same thickness?

R

Research activity

For the lavatory pans that we use in this country find out the following information:

- where they are made;
- what materials are used;
- where the materials come from;
- how they are made;
- what else is needed for them to work properly.

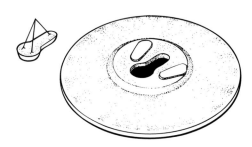

◨ *This slab covers a latrine hole and improves hygiene*

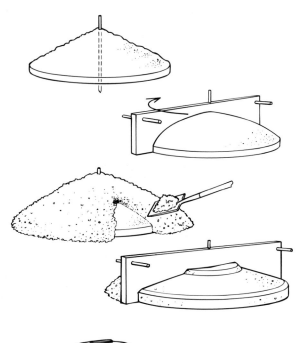

◨ *The slabs are made using a concrete mix of 3 parts cement, 6 parts clean sand and 4 parts ballast*

Economic mass production

How do these slabs improve the quality of people's lives other than by raising standards of hygiene? ApproTEC has redesigned the production equipment and techniques to allow for simple and high quality mass production of the slabs. They can be manufactured using locally made wooden and metal tools rather than expensive imported kit. This means local entrepreneurs who have the equivalent of about $100 can set up a small business making the slabs. This in turn provides local people with work using tools and materials they are already familiar with. The new business also creates work for other local industries making tools and providing materials.

Questions

1 **a** Draw a flow diagram describing the production of the latrine covers.

  b For each material or piece of equipment used, note where it comes from and/or how it is made.

  c Use this information to decide if the ApproTEC latrine cover is appropriate technology. Look at page 93 for extra help.

ApproTEC

Members of ApproTEC were originally with Action Aid, one of the largest aid agencies working in Africa. Since 1990, they have developed a separate organization which promotes the creation of sustainable employment and economic growth. To do this they develop and promote technologies appropriate for small-scale enterprises in Kenya and the region.

They believe that developing opportunities for private sector entrepreneurs is the best form of development work. This is done by encouraging small, labour-intensive industries to fill new market niches. They have devised a programme of research, training, engineering design and development as well as specific enterprises and technologies.

In addition to other schemes, ApproTEC has trained and supervised local people enabling them to install more than 30 000 domed pit-latrine slabs. It has also trained aid agency staff in setting up the manufacture of domed pit-latrine slabs in refugee camp situations.

The key factor in all its work is that it does not impose something which is unrealistic and unattainable, but provides a way for communities to develop a self-supporting local economy that is appropriate to the environment and culture.

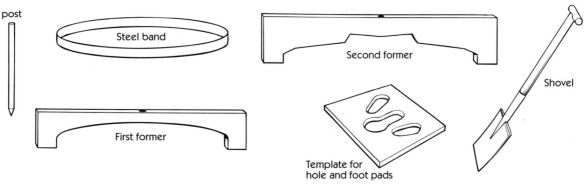

Tools needed to make the ApproTEC latrine slab

House of card

Cardboard sheets and chopsticks – a recipe for cardboard furniture! Olivier Leblois, a French architect, designed a range of furniture including chairs, tables and shelving. In discussions with Quart de Poil, which specializes in developing modern furniture and in particular cardboard products, the company suggested he explore the possibility of developing his designs using corrugated cardboard rather than wood.

Designing for performance and manufacture

Leblois worked with Rexam Plc (formerly Bowaters), a cardboard manufacturer in the UK. Rexam's role was to translate his drawings for an adult's chair into card prototypes. Rexam was able to do this because the company knew a lot about the properties of cardboard, for example that it is stiffer than steel of the same weight per area. It is also expert in manufacturing with cardboard so it was able to develop designs that could be made.

It took about one month to get it right and involved testing four prototypes for comfort, stability and durability. At each stage of prototyping a flat plan was produced for the component parts. These plans were then used as a pattern for the pieces which were cut by hand before being assembled.

P **Pause for thought**

Is cardboard really strong enough for furniture?

For large-scale manufacture a computerized tracing table was used to cut each part from large sheets of card.

In developing a flat plan that could be manufactured Rexam had to take the following into account:

- no acute angles;
- joining pieces must maintain their strength;
- effective use of materials by getting as many parts as possible out of one piece of board;
- maintaining the board's inherent strengths by maximizing the use of the direction of the corrugations, just like with wood or textiles where pattern pieces must be cut with the grain.

This cardboard furniture has been overprinted

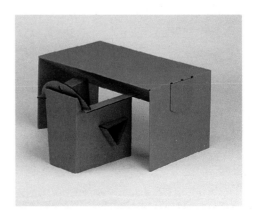

Key features

The range is not intended to replace people's existing furniture but has been developed as quality temporary or 'short-life' items. Leblois and Rexam believe that the furniture has a lifespan of around eight months if used every day and kept in a dry environment.

The product has several important selling points:

- flat-pack;
- easy to put together – no tools required;
- can be overprinted or painted by hand for a personal touch;
- cheap;
- recyclable;
- wood pulp for producing cardboard has come from farmed trees.

Questions

Use the plans shown to make a small-scale model of Leblois' cardboard chair.

Selling the furniture

Quart de Poil has patented and marketed Leblois' designs. The company has identified three major markets.

- People who buy through retailers such as Galleries Lafayette in France and Liberty in England where the furniture is promoted as additions to existing items rather than replacements. As a result it is never sold in high street furniture shops.

- Advertising agencies who use the furniture to promote companies or products by having the furniture overprinted with logos or product names as appropriate. For instance, one promotional campaign used the adult chair overprinted with the company's name and address to send to all its existing and potential clients. This is the largest market.

- The furniture has also been used extensively at exhibitions and trade shows.

Research activity

Investigate the availability of card furniture in your area. Find out:

- where you can buy it;
- what sorts of furniture you can buy;
- how much it costs.

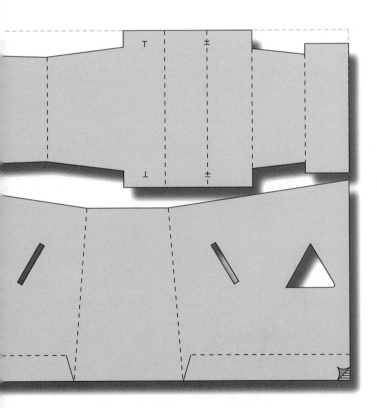

All the parts for a chair from a single sheet

Manufacturing products

Manufacturing a stapler

A stapler is an everyday object. You probably use one at school and at home. The way a stapler works requires all the parts to be made accurately and assembled carefully. The diagram shows all the different parts that make up a stapler and how each one is manufactured.

R

Research activity

Choose another everyday object, for example a paper punch, ball-point pen or tin opener, and find out how each part is manufactured and the whole product assembled. Present your findings as an exploded diagram, with labels.

SPARE PARTS CHART FOR STAPLING MACHINES

WHEN ORDERING PLEASE STATE MODEL AND PART No. →

	PART No.	DESCRIPTION	PRICE
1	D03522	BODY	
2	D03524	FEEDER 'N'	
3	D03525	STAPLE STEADY	
4	A03540	COVER ASSY 'L/GREY'	
5	D03531	COVER 'LIGHT GREY'	
6	D03533	COVER BRACKET	
7	D03536	PLUNGER BLADE 'N'	
8	D06341	COVER KNOB 'GREY'	
9	A03560	CATCH SPRING ASSY	
10	D03555	CATCH COVER	
11	D06024	LIFT SPRING	
12	D06043	HINGE PIN	
13	D06344	BASE PAD	
14	D06257	BASE 'GREY'	
15	A06345/AA	BASE ASSY 'GREY'	
16	D07779	ANVIL RIVET	
17	D07972/AA	ANVIL 'N'	
18	D08674	FEEDER SPRING	
19	D08962	ANVIL SPRING	
20	D81410	RIVET	

☐ – ASSEMBLY OF PARTS

◯ – SINGLE PART

SERIES 1500
REXEL STELLA STAPLER 'GREY/GREY'

M/C No. A01588
DRG. No. D10702
ISSUE No. 0190

Focused case studies

3

Manufacturing methods

The method of manufacturing chosen will depend on the number of items to be produced. Some of these are summarized in the table.

Methods of manufacture

Individual items	Small batch	Quantity manufacture
sand-casting – wooden patterns	sand-casting – metal patterns	extrusion
manual machining	shell moulding	injection moulding
steam bending timber	vacuum forming	die casting
laminating timber	blow moulding	stamping and pressing
manual welding, soldering and brazing	drape forming	computer-controlled machining
		robot welding, soldering and brazing

Note that with the development of designs specifically for computer-assisted manufacture, small batches and one-offs are sometimes produced 'as needed' using computer-controlled machines.

Manufacturing systems

Manufacturers need to organize their methods of production into manufacturing systems. Two important systems are explained below.

Cellular manufacturing

This system uses small teams of people called cells. Each cell is responsible for the manufacture of a complete product. Each person in the cell is multi-skilled and does a number of different jobs. This makes the work more interesting for each cell member.

The cell is responsible for quality; there are no separate inspection stages. It operates in a small physical area and there are no long delays between different processes – wasted time is reduced, adding value to the product.

The overall effect of the cell system is greater efficiency, increased output and improved quality.

Just-in-time manufacturing

This system is organized to meet immediate market demand. Stocks of materials and parts are delivered 'just in time' to make what customers have ordered, so there are no long-term storage requirements. All processes are carried out 'just in time' to meet the delivery dates so there is no build-up of finished products.

Cells can operate on a just-in-time basis.

Both the cellular and just-in-time systems rely heavily on the initiative of the production workers. They have to make on-the-spot decisions about what materials and parts to order and when to order them, as well as how to process them once they have been delivered.

Research activity

Find out if there are any manufacturing companies in your area and make a list of them. Find out which manufacturing methods each one uses, and which manufacturing systems.

Energy to make things work

The power rating of an appliance tells you how much energy is needed to make it work each second. The table on the right shows the power ratings of some appliances people use every day.

Such amounts of energy have not always been available, as the panel below shows.

Appliance	Power rating
radio	10 watts
microwave	650 watts
electric iron	1000 watts
2-bar electric fire	2000 watts
washing maching	2500 watts
family car engine (under braking load)	37 500–45 000 watts

1000 watts = 1 kilowatt.   *1 watt = 1 joule per second.*

	Time	Power	Comment
Life without engines	prehistoric times	140 watts for a strong person	Even with simple machines and using animals, life without engines is very hard work
Wind and water power; the first engines	7th century for waterwheels in Europe after the Roman Empire, 12th century for windmills in Europe	primitive waterwheels, 300/400 watts late 18th century windmills, 2–3 kilowatts	These low-power sources are suitable for monotonous tasks like grinding corn and operating forge hammers Water power requires a suitable river and the power is only available at or very near the waterwheel Wind power requires a windy place. The power is only available at or very near the windmill and when wind is blowing
Atmospheric engines	18th century	Newcomen's engine, 3.5 kilowatts Watt's engine, 7–8 kilowatts	Can operate wherever engine can be built and fuel transported Power only available at or near the engine site

	Time	Power	Comment
True steam engines	19th century	By the 1880s, 5000 kilowatts	Can operate wherever engine can be built and fuel transported Power only available near the engine site
Generating electricity	late 19th century	100 kilowatt generators used for street lighting in New York in 1882	Principles of the dynamo and electric motor understood by mid-19th century Scene set for generating and transmitting electricity which can be used to power machines that work by electric motors
Generating electricity using fossil fuel	20th century	Modern power station in the UK generates 500–1000 megawatts	Contributes to acid rain unless advanced technologies used Contributes to greenhouse effect
Generating electricity using nuclear fuel	late 20th century	Modern power station in UK generates 900 megawatts	Does not contribute to acid rain or greenhouse effect but concern over safety of power stations and disposal of radioactive waste
Generating electricity using renewable energy sources	late 20th century	Modern windmill generates 3–4 megawatts Tidal barrier generates 240 megawatts	Does not contribute to acid rain or greenhouse effect but concern over other types of environmental impact

Focused case studies

Identifying needs and likes

You can revise strategies for identifying needs and likes from key stage 3 by thinking about these people who visited the dentist one Tuesday morning between 9.00 and 10.30.

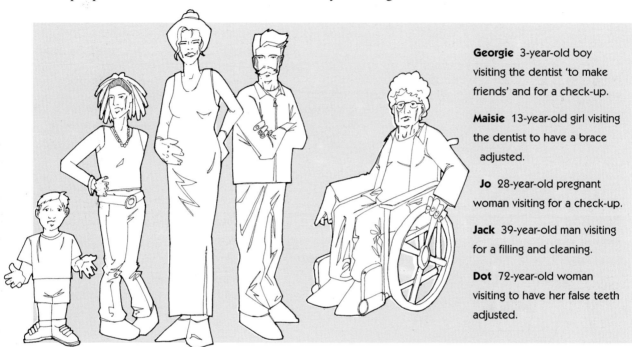

Georgie 3-year-old boy visiting the dentist 'to make friends' and for a check-up.

Maisie 13-year-old girl visiting the dentist to have a brace adjusted.

Jo 28-year-old pregnant woman visiting for a check-up.

Jack 39-year-old man visiting for a filling and cleaning.

Dot 72-year-old woman visiting to have her false teeth adjusted.

Thinking about what people might need

The people visiting the dentist will have different needs and likes. You can try thinking about these by using the PIES approach. PIES stands for **p**hysical, **i**ntellectual, **e**motional and **s**ocial. Each of these words describes a type of need that can be met by products that have been designed and made.

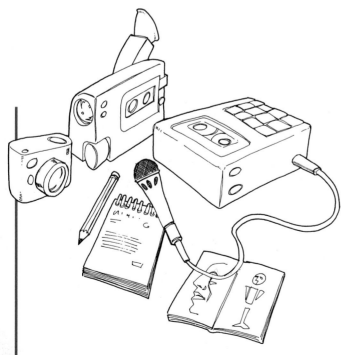

Observing people

You can find out a lot about people's needs and likes by watching them. It is important to record your observations in a way that doesn't affect what the people are doing. The illustration shows several different recording methods. Can you explain which ones are suitable for use in a dentist's waiting room?

Asking questions

You can find out about people's preferences by talking to them and asking questions. This is sometimes called **interviewing**. It is different from using a questionnaire as you only interview a few people. It is important to ask the right sorts of questions. To find out what each of these people want from their dentist's waiting room you would probably need to ask them different questions.

Using books and magazines

Sometimes you need to find out something by looking things up in books and magazines. Some magazines might tell you about the preferences of people using dentists' waiting rooms. Where would you find these magazines? Some books might tell you about the rules and regulations governing dentists' waiting rooms. Where would you find these books?

Image boards

You can make a collection of pictures of things that a person or group of people might like, places they might go, activities they might do. This is called an **image board**. An image board for Georgie would look very different from one for Maisie. Making image boards will help you understand what different people might like. It may also help you understand the style of products that would appeal to different people. For example Jo and Dot may both wear wristwatches but they will almost certainly look very different.

Q **Questions**

Here is the beginning of an image board for Jo.

1 What does it tell you about her?

2 What other images could be added to give a fuller picture?

Play Home children food family Health clothes work friends

Who is Jo?

Using questionnaires

A questionnaire is a carefully designed set of questions. It is often used by businesses to find out what different groups of people like or would buy. A questionnaire will usually try to get information about the sort of person who is answering it – their occupation, how much they earn and so on. This information enables businesses to provide goods and services that people want at a price they are prepared to pay. It also shows where and when these products could be sold and how best they might be advertised.

Designing your own questionnaire

You need to be clear on what you are trying to find out. Target your questions to obtain the information you want. Avoid leading questions that suggest the answer. Avoid questions that don't discriminate, such as 'Do you like sunny days?'. Everyone always answers yes!

Sometimes you will use the questionnaire in face-to-face questioning when you record people's answers. At other times people will fill it in on their own and return it to you. In this second case it is particularly important that the meanings of the questions are clear as you won't be there to explain them.

Advice on writing questionnaire questions is given in the panel on the right.

Questions

Notice how newspapers and magazines use so-called questionnaires to attract the readers' interest rather than provide useful information.
What sort of information do these questionnaires reveal to the readers?

Questionnaire question guide

- Use closed questions. These require a yes or no answer or give people a choice of answers.

- Make it easy to fill in the answers. Use tick boxes where possible.

- Each question should be short and simple.

- Use words people will understand.

- Write questions which only have one meaning.

- Each question should ask only one thing at a time.

- A scaled choice of answers is a good way to find out people's attitudes.

What sample size should I use?

It is important to present your questionnaire to as many people as possible. This will give you a large number of responses from which you can draw reasonable conclusions. A hundred responses would be an ideal number, but this would be a huge task for one researcher. If the research is shared amongst a group of people the task becomes manageable both in terms of collecting responses and collating data. If each member of a class of 20 students took responsibility for 5 questionnaires the sample size would be 100.

Collating the results

Once you have the returned questionnaires you will need to analyse the information. Here's how to do it.

- Draw up a summary results table or tally sheet of the possible answers to each question.

- Count how many of each possible answer you got for each question and write this in the table or on the tally sheet.

When you have done this for each question on each questionnaire the table is complete and you can begin to think about what the results mean. You will find that putting the information into a database or spreadsheet may help you collate it more quickly.

Using spreadsheets and databases

The database will organize the information so that it is easily accessible and can be displayed clearly. The database can be 'interrogated' for statistical information and thus provide a picture of user needs and likes.

Statistical information from the database can be put into a spreadsheet. The spreadsheet displays the information as rows and columns of numbers. You can analyse the information in a variety of ways and present your findings in graphical forms like pie charts and bar graphs.

Here is an example of a survey of lighting. The completed questionnaires provided information on lighting in students' homes. This information was recorded on a tally sheet. It was then analysed using a spreadsheet to identify user preferences. The results were presented visually.

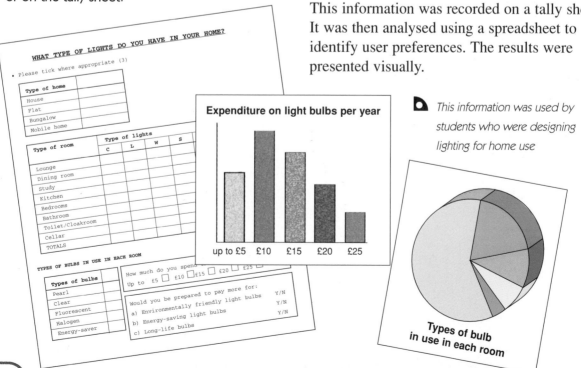

This information was used by students who were designing lighting for home use

Design briefs

A design brief is a short statement which describes some or all of the following:

- the sort of product that is to be made and its purpose;
- who will use it;
- where it will be used;
- where it might be sold.

An **open** brief provides general guidelines and offers opportunity for a wide range of possible outcomes. A **closed** brief is more specific and detailed in its requirements.

Here are examples of open and closed briefs for two lines of interest.

Lighting

Open design brief: Design a lighting system for a new restaurant called the Garden House. It will seat up to 80 people at 20 tables. It has 'country gardens' as its theme and it aims to attract shoppers and families during the day.

Closed design brief: Design a ceiling-mounted, rise-and-fall light that casts a warm glow on a round table of diameter 2 m. It is for a restaurant that aims to attract couples and small parties for 'special occasion' evening meals.

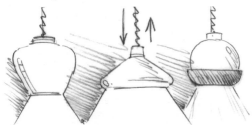

The open brief provides the designer with freedom to explore and create a wide range of possible lighting effects.

The closed brief provides opportunity for different solutions, but the nature of the product is more clearly defined so the range of possible outcomes is limited. A particular lighting effect is required and there are only a few ways this can be achieved.

Toys

Open design brief: Design toys based on the story of *Goldilocks and the Three Bears* for children of primary school age. They are to be sold in a wide range of outlets: specialist toy shops, department stores and Woolworths.

Closed design brief: Design a wooden jigsaw for infant school children based on the *Goldilocks and the Three Bears* story. It is to be sold in craft shops that specialize in hand-made toys.

A range of toys is possible from the open brief including traditional dolls, mechanical automata and computer games.

In the closed brief the product and construction material are specified and the end user is more clearly identified. This provides a more detailed picture of what is required.

Specifying the product

You will need to develop the design brief into a **performance specification**. This will provide a list of criteria against which you can assess your design as it develops.

The performance specification will always:

- describe what the product has to do;
- describe what the product should look like;
- state any other requirements that need to be met.

For example:

- how it should work;
- how much it should cost to manufacture;
- possible production levels – one-off or batch production;
- what materials it should be made from;
- what energy source should be used if it needs to be powered;
- ergonomic requirements related to end user;
- legal requirements to be met in its development and use;
- environmental considerations and requirements.

Here are two examples of performance specifications and products that meet their requirements.

Infant school bench hook specification

What it should do:

- hold secure a range of round/square section materials (timber/plastic) to allow for safe cutting;
- be fixed to classroom table/desk and removed easily when not in use/needed.

What it should look like:

- a robust, but friendly tool;
- attract user because it looks easy to operate;
- line, shape and colour scheme to compliment general classroom environment.

Other requirements:

- must be safe and easy to use by young children and inexperienced adults;
- must not endanger non-users in general classroom situation;
- must be adjustable and accommodate a range of materials up to 25 mm × 25 mm;
- clamp/holding device should be easy to lock and release. Young children should be able to hold material secure with minimum force;
- all components should be attached securely so that they do not become separated and lost.

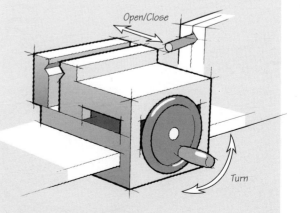

Body adornment specification

What it has to do:

● be suitable for male/female wear for a party or celebration;

● be part of a range of low-cost fashion jewellery which can be mixed and matched to coordinate with evening/party wear;

● meet the aesthetic needs of a young person aged 17–25.

What it should look like:

● be based on geometric forms with reference to Egyptian art and design;

● reflect mood/style of fun-loving young person.

Other requirements:

● must be easy to manufacture for batch production;

● should use low-cost materials, recycled if possible;

● should be seen as an evolutionary range of jewellery with opportunity for regular style changes and special/limited edition products (like a Swatch Watch).

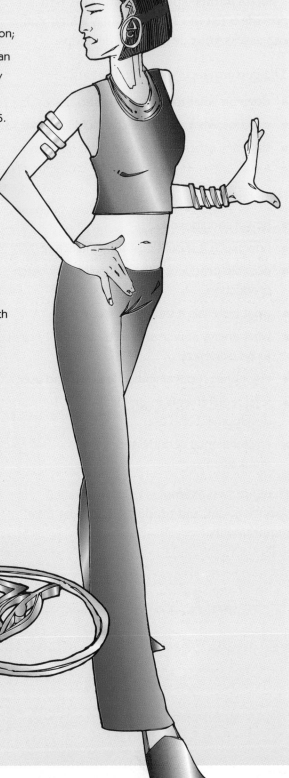

Generating design ideas

Brainstorming

You probably did some brainstorming at key stage 3. Here is a reminder.

Brainstorming is:

- a process for getting ideas out of your head!
- a process for getting ideas you didn't know you had!
- a process which uses questions and associations and links ideas to actions;
- a process you can do on your own, but it is usually better in a group.

Brainstorming an idea can help you to identify a wider range of options for your designing and making and to work out how best to develop these ideas.

How to brainstorm

- First state the problem or need.
- Record every idea suggested as words, phrases or pictures.
- Produce as many ideas as possible.
- Don't make judgements until the brainstorming pattern is complete.

- Allow enough time for new and diverse ideas to emerge, but agree a time limit so that ideas remain fresh.
- Sort out ideas by considering which are unrealistic, inappropriate and unachievable and removing them. What is left will give you a focus for action.

What can I use for this?

By asking this question you can identify design options. You can give each possibility a yes/no verdict based on specific criteria – availability, cost, effectiveness and feasibility. You can refine the remaining options using similar criteria until you are left with a 'best' solution. Here is an example.

Young plants planted outdoors at the end of Spring are often destroyed by unexpected late frosts. What can be done to protect them? The obvious solution to this problem is a protective covering, but what sort of covering? The design of the protective covering is the focus for brainstorming.

This brainstorming session gave us a best solution for a plant cover – a design using a flexible sheet as a shell form rendered rigid by the use of ties, and held in place by being bolted to metal stakes. Note that the brainstormers used the Structural Element and Joints and Connectors Chooser Charts.

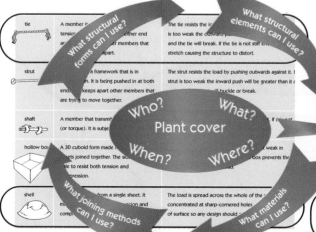

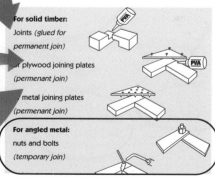

What can I use this for?

This is the sort of brainstorming that you use when you have some technical capabilities and aren't sure how to use them.

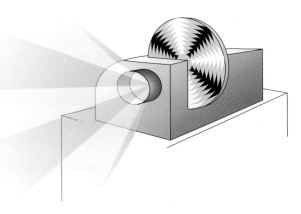

Imagine that you know how to control movement with an electric motor and a simple switching system. This allows you to turn the motor on and off and make the motor output shaft rotate clockwise and anticlockwise. You can also control the speed of rotation: fast, medium and slow. You can use brainstorming to find something useful to do with this knowledge.

Here is an example. Notice how the brainstormers have used the PIES approach within their brainstorming.

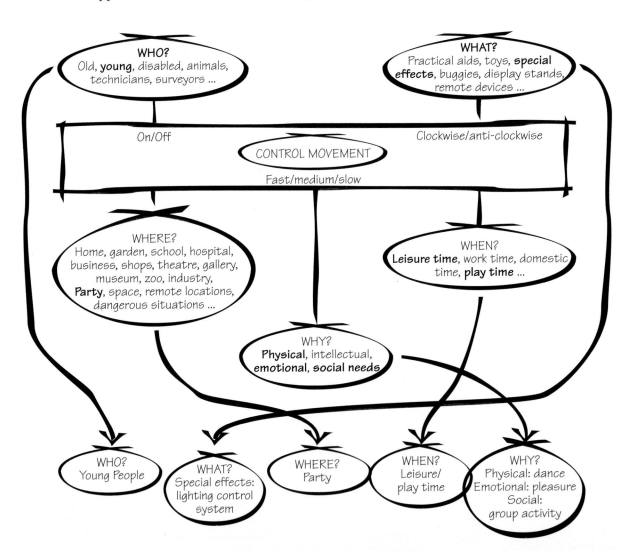

Attribute analysis

You may have used attribute analysis at key stage 3. Designers and engineers use it to help them produce new designs for familiar products.

Here is an attribute analysis table for a piece of body adornment. The headings describe attributes which will affect the final design. You can read across the columns and combine different words from each column to create new designs. Some combinations will be totally inappropriate, while others will offer viable design ideas. The medium-cost metal bangle for special occasions for young people aged 19–25 years seems a viable design idea. The high-cost card bracelet with the Velcro fastening seems unlikely to appeal to people over 65.

Type	Based on	Materials used	Fastenings used	Price	When worn	Typical wearer
brooch	natural form	wood	clips	low	**special occasion**	male
badge	**geometric form**	**metal**	rings	**medium**	day-time	female
necklace	**abstract form**	plastic	studs	**high**	evening	<12
pendant		paper	pins		**every day**	13–18
bangle		**card**	buckles			**19–25**
bracelet			**Velcro**			25–35
headpiece						35–45
hat-pin						45–55
hair-slide						55–65
earring						**65+**
ear stud						
cufflinks						
tie-pin						

Identifying needs and likes

You can revise strategies for identifying needs and likes from key stage 3 by thinking about these people who visited the dentist one Tuesday morning between 9.00 and 10.30.

Observational drawing	Resulting design
These drawings of chairs provided a store cupboard of ideas for designing the child's wooden chair. Foam arm chair Wooden arm chair frame Box stool	Child's wood chair
These drawings of brick patterns helped with the decoration of the metal box.	Metal box
These drawings of leaves helped with the design of a child's wooden puzzle. Holly Lime Ivy Horse Chestnut	Child's wooden puzzle
These drawings of bridges helped with the design of the cantilever light. Motorway bridge Cantilever bridge Beam bridge	Cantilever lamp

Investigative drawing

You can investigate the way something works by doing careful drawings that try to explain *how* it works. Here's how to do it.

- First find out how it works by using it and looking at it quickly.

- Write down what you have to do to make it work and what you think might be happening when it works.

- Then investigate how it works by looking more closely. Use a hand-lens for close-up views. Look inside and if necessary undo parts to get a good view.

- Draw the parts you can see and add notes and other drawings to show what the different parts do.

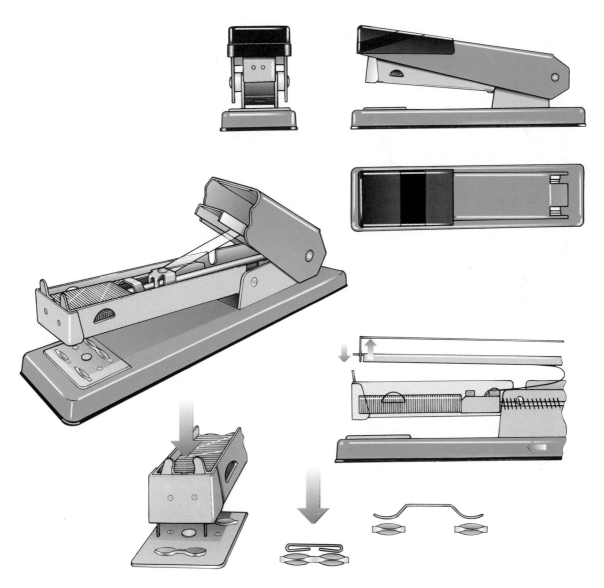

An investigation of the way a stapler works produced this series of drawings. By looking closely at the various components and trying to understand how they fit together you can gain a better understanding of how the stapler works

Modelling

It is often difficult to imagine what a design idea will look like or how it will work. Modelling your design ideas gives you something to look at, think about and test.

Modelling will help you:

- clarify and develop your design ideas;
- evaluate your design ideas;
- share your design ideas with others.

Modelling appearance

There are many modelling techniques, some of which you will have used at key stage 3. Here are examples of the way modelling techniques have been used to develop the designs for a piece of jewellery.

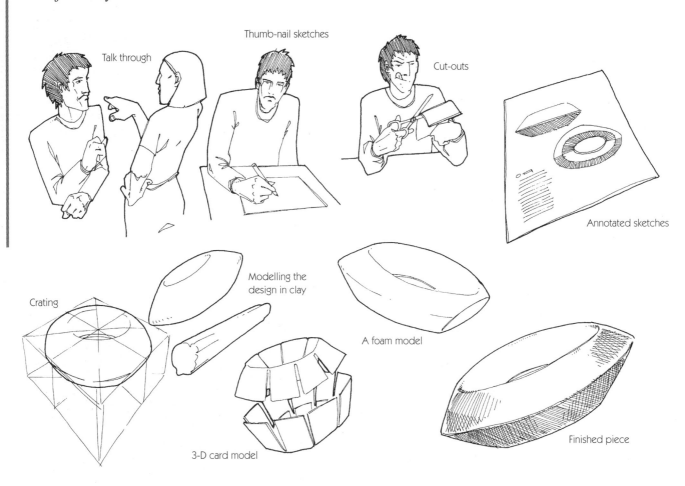

Talk through

Thumb-nail sketches

Cut-outs

Annotated sketches

Crating

Modelling the design in clay

A foam model

3-D card model

Finished piece

◗ Modelling appearance is a useful process for assessing design ideas for development into real products

MFRT1   MFRT2   LIRT1

Modelling product performance

Modelling not only describes the way a product will look, it can also describe the way it will fit together and how it will work. You can use a range of modelling techniques to develop design ideas about product performance. Talking about the design, thumbnail sketches and detailed drawings will help you model how a product might work. You will need to use a more sophisticated approach for products that involve some form of movement or control. Some of these modelling techniques are shown for developing designs for adjustable seats.

Whenever you design products that people will use you will need to think about sizes and shapes (**anthropometrics**) and movements (**ergonomics**). Tables of data are available and you can use this information to make your product easier to use. All the chair designs shown here would need to be informed using anthropometric and ergonomic information.

You can assess some of the effects of a product on a place where it will be used by modelling both the product and the place. For example a scale model of an office chair placed in a scale model of an office in which it will be used provides an opportunity to assess its appropriateness and performance in that environment. You could find out if it looks right, if it takes up too much room, if it fits in with the other furniture and so on.

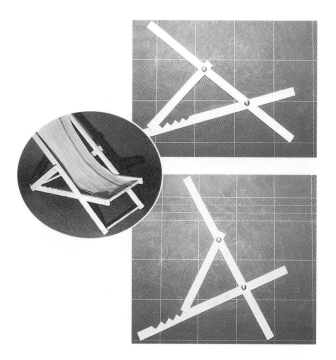

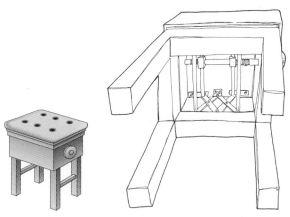

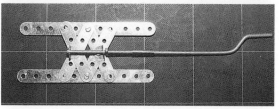

Modelling adjustable seats

Modelling appearance and form with computers

If you use computers properly they can help you model your design ideas so that you can explore many more possibilities than if you were working just with pencil and paper. There are several ways to start using the computer as these examples of design show.

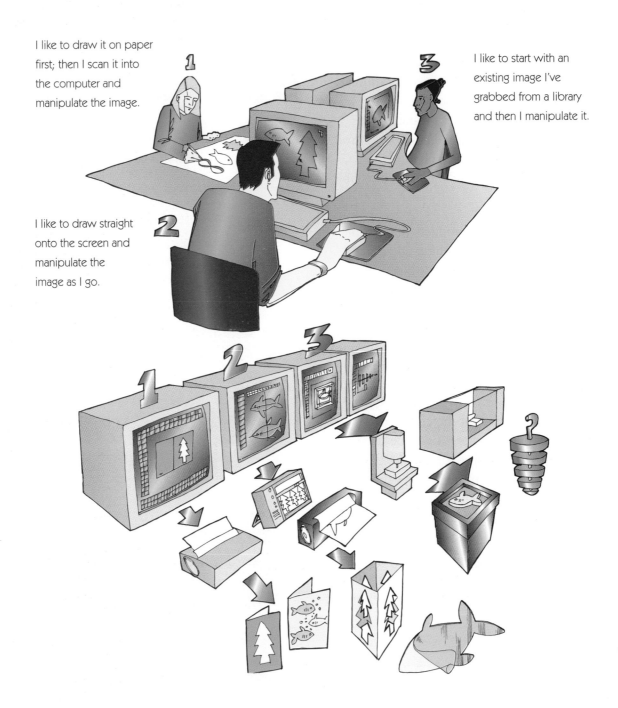

I like to draw it on paper first; then I scan it into the computer and manipulate the image.

I like to start with an existing image I've grabbed from a library and then I manipulate it.

I like to draw straight onto the screen and manipulate the image as I go.

Modelling function with computers

You can use a computer to model the way a product might work. You can make changes to the design 'on-screen' and see how this affects the performance.

The design for the nursery light involves both mechanical and electrical systems. Modelling these systems on a computer will help you get the technical details right so that they work well.

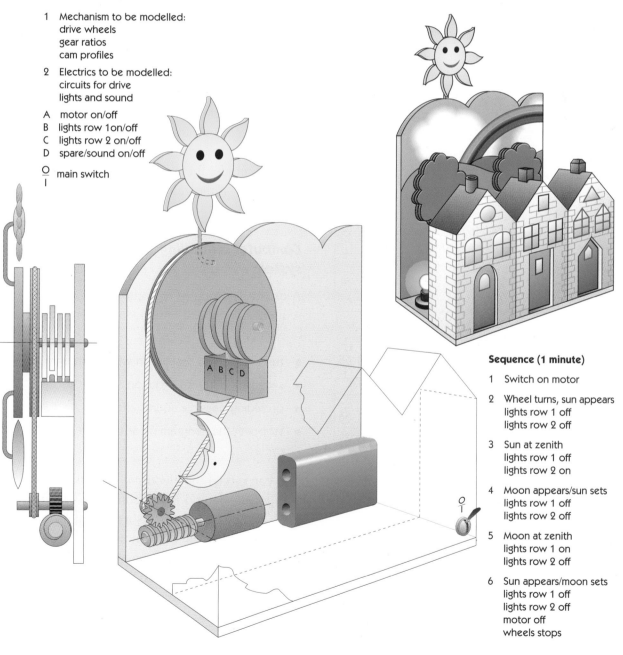

1 Mechanism to be modelled:
drive wheels
gear ratios
cam profiles

2 Electrics to be modelled:
circuits for drive
lights and sound

A motor on/off
B lights row 1 on/off
C lights row 2 on/off
D spare/sound on/off

O
— main switch
I

Sequence (1 minute)

1 Switch on motor

2 Wheel turns, sun appears
lights row 1 off
lights row 2 off

3 Sun at zenith
lights row 1 off
lights row 2 on

4 Moon appears/sun sets
lights row 1 off
lights row 2 off

5 Moon at zenith
lights row 1 on
lights row 2 off

6 Sun appears/moon sets
lights row 1 off
lights row 2 off
motor off
wheels stops

Nursery light automaton: design concept

Strategies – modelling

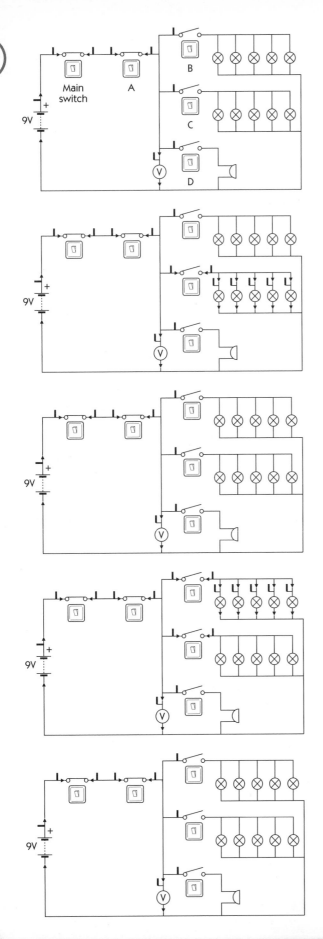

Main switch

A

B

C

D

9V

Computer modelling the mechanical system

The movement of the sun and moon wheel is driven by an electric motor. The speed is controlled by the worm and wheel and the belt and pulley. It is important that the speed is suitable for the nursery light: the sun and moon should each be visible for about one minute. You can model the effect of pulley and wheel size in order to get the speed you need.

The light switches are operated by a series of cams. It is important that the lights switch on and off according to the time of day as revealed by the position of the sun and moon. You can model the effect of the shape and orientation of the cams to get the switching you need.

Computer modelling the electrical system

A circuit is required to control the lights. A sequence is needed which starts a programme of events:

● turning the sun and moon wheel through 360°;

● switching different lights on and off;

● providing possible sound effects;

● stopping the programme at the end of one cycle.

You can model the electrical circuits using components stored in the computer's memory. Each component performs a particular function which can be demonstrated. You can model a working circuit on-screen and adjust it to give the performance you want. Then when you build the real circuit you can be sure it will work.

◧ Computer modelling of circuit for nursery toy.

Applying science

Checking on your choice of material

Strength is one of the properties of all solid materials. If the strength of a material is high then it will be difficult to break. If the strength of a material is low then it will be easy to break. You can find out how strong some materials are by looking them up in the Materials Chooser Charts on pages 188–194. The chart will tell you about the **tensile strength** of the materials, that is how hard you have to pull on them before they break. The chart will also help you put materials in an order of strength.

If a part of your design isn't strong enough you can do one of two things:

- change the design so that the cross-sectional area of the part is greater – if there is more material there it will be stronger; or

- make the part from a stronger material.

You may find that a material is strong enough but too elastic – it doesn't break under the load but it does bend. This elastic property of a material is often measured as Young's modulus of elasticity. You can find out how elastic materials are by looking them up in the Materials Chooser Charts.

If a part of your design bends too much you can do one of two things:

- change the design so that the part has a cross-sectional shape that is less bendy; or

- use a material that is less elastic.

You can find out about cross-sectional shapes and stiffness on page 177.

Plastic	acrylic	PVC	nylon	polystyrene
Durability	●●●●	●●●●	●●●●●	●●●●●
Softening point/°C	85–115	70–80	230	80–105
Relative price/£	●●●	●●	●●	●●
Ease of sourcing	●●●●	●●●●	●●	●●●●
Hardness (how difficult it is to scratch)	●●●●	●●	●●●	●●
Strength (how difficult it is to break)	●●●	●●●●	●●●●●	●●●●
Density (how heavy it is)	●●●	●●	●●●	●●
Modulus of elasticity (how difficult it is to stretch)	●●	●●●	●	●
School uses	🔧	📦	✏️🤖	📦🗑️
Ease of hand-working	○○○	○○○○	○○○	○○○○○
Ease of processing	○○○○○	○○○○○	○○○	○○○○○
Suitable processes	line bending	vacuum forming	turning component	vacuum forming

Most materials react with their surroundings in some way. Some react only slightly and it takes a long time for them to corrode or rot. Others decay quite rapidly. The time taken will depend on the material and the severity of the conditions. You can check on the durability of materials by looking them up in the Materials Chooser Charts.

You can improve the durability of your design in two ways:

- use a finish on the material which protects it from the surroundings; or

- choose a more durable material.

Strategies – applying science

Getting surface effects on metals

Electroplating

You can plate copper onto brass by making the brass the negative electrode in the electrolysis of a copper sulphate solution. The copper formed at this electrode sticks to the brass as copper-plating. By masking the brass with permanent marker you can create interesting patterns.

Anodizing

You can produce a range of attractive colours on titanium by using it as the positive electrode in the electrolysis of Lucosade. The oxygen gas liberated at this electrode combines with the titanium to give different coloured oxides.

Surface colouring

You can produce an attractive dull, mottled green colour on copper by keeping the metal object in sawdust that has been treated with ammonium chloride and ammonium carbonate solution. The treatment should last about 14 days.

You can produce a beautiful deep red sheen on copper by heating it to red heat and then plunging it into rapidly boiling water. This is only suitable for items made from a single piece of copper. Can you explain why?

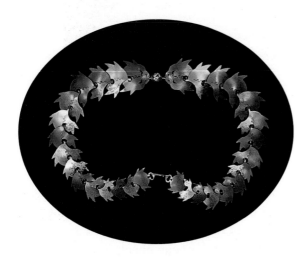

The different colours in this titanium necklace have been produced by anodizing

This bronze bowl has been coloured green by the sawdust – ammonium salt solution treatment

Thinking about forces

How much force will I need?

Here are two examples showing how to measure the force needed for simple tasks.

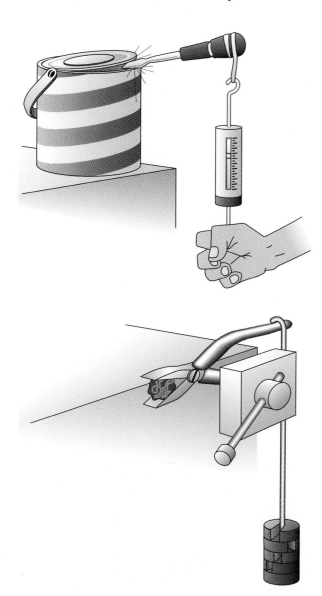

▶ Where else might you need to take measurements?

How much will it lift?

You can measure how much a motor will lift by using the method shown below.

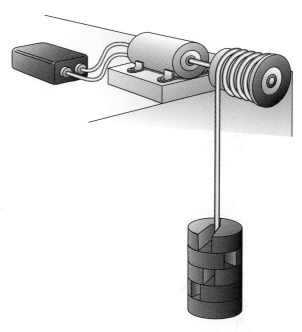

▶ What will happen when there is enough force to crack the nut?

How can I get it to balance?

You can calculate how much force you need to get things to balance by using the principle of moments (see page 179).

Systems thinking

During your key stage 3 work you may have been introduced to systems thinking. You can use this to help you understand complex products. Here is a summary of the important ideas using a music centre as the example.

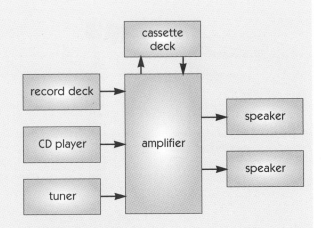

The controls on the music centre make up the user interface. It is important to design user interfaces that are easy to understand and simple to use.

This music centre can be thought about in terms of the subsystems that make it up. You can show the subsystems by drawing a systems diagram. It doesn't look like the music centre but it does help you understand how it works.

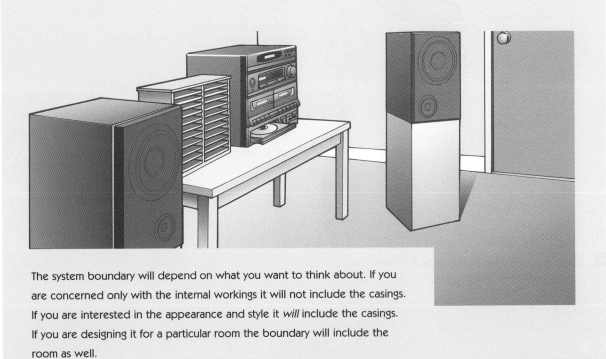

The system boundary will depend on what you want to think about. If you are concerned only with the internal workings it will not include the casings. If you are interested in the appearance and style it *will* include the casings. If you are designing it for a particular room the boundary will include the room as well.

System interfaces

The picture shows testing equipment to measure the strength of sewing threads. You can think about this equipment in terms of five different systems. Thinking like this is useful because it can help you decide on the way the systems are linked to one another. These links are the **interfaces** between the different systems. For example, the mechanism for holding the thread must fit onto the main body of the test equipment
in a way that enables it to move easily. The structure of the test equipment must be made from parts that are strong enough to resist the forces used to test the thread.

Thinking like this will help you design a complex product so that each system works well with the other systems in the product.

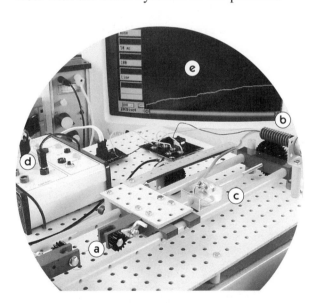

a Mechanical system for gripping the sample.

b Electrical system using a motor to apply the pulling force and a strain gauge to measure the deflection.

c Structural system for supporting the mechanical and electrical systems.

d Data capture system of the interface and computer that records the force applied and the deflection of the thread.

e Data presentation system of the software that plots force applied against deflection.

Feedback and control

Control systems need information so that they can respond to changes. This information is called **feedback**. Systems with feedback are called closed-loop systems. A system without feedback is called an open-loop system. The drawings below show both sorts of system.

There is no feedback to tell the tap that the sink is full

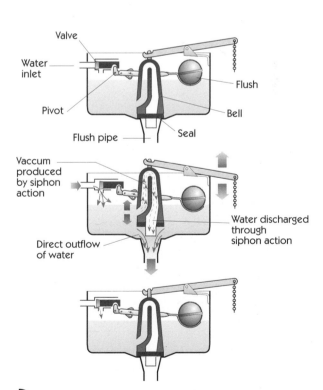

The level of the water is monitored by the float. The arm attached to the float controls a valve which controls the flow of water into the tank

Planning

Flow charts and Gantt charts

You can use flow charts and Gantt charts to help you plan your way through a Capability Task. In year 11 you may spend two whole terms on a single Capability Task as part of your GCSE assessment. It will be important to ensure that school holidays, public holidays, sports days, etc. don't upset your plans.

You can use the headings in the flow chart shown opposite to get the order of the task right. Once you have the order right you can use a Gantt chart to think about how long each part should take and to make sure that you get the task done on time. A Gantt chart will give you an overview of the whole task, showing both what needs to be done and when it should be done.

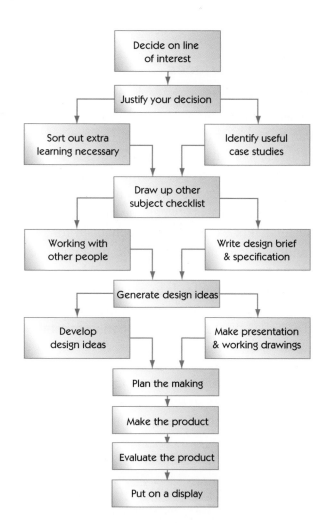

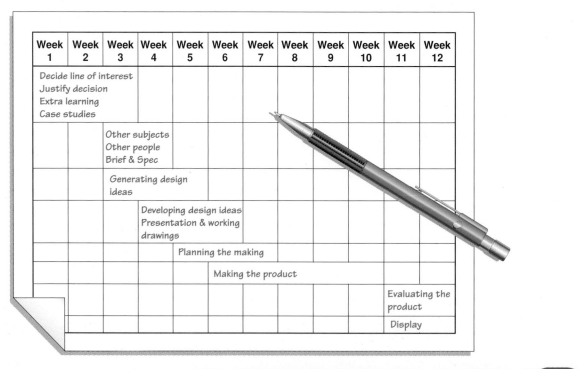

Evaluating

User trip

The simplest way to evaluate a product is to take a user trip. This involves using the product and asking a few basic questions:

- Is it easy or convenient to use?
- Does it do what it is supposed to do?
- Do I like it?
- Would I want to own or continue to use it?

The local authority has replaced the traditional fold-up deckchairs in the park with stacking plastic chairs which are cheaper and easier to maintain. However, will the public support this change? The members of our user group, Georgie aged 3, Maisie aged 13, Jo aged 28, Jack aged 39 and Dot aged 72, have been asked for their views and are taking a user trip. Their thoughts about the chair will be expressed when they respond to the user trip questions.

Winners and losers

The outcomes of design and technology will provide benefits for some and disadvantages for others. Designing and making a product will affect a lot of people directly and indirectly.

Maisie's brother, Wayne, has worked hard doing odd jobs around the house for his parents to raise enough money to buy himself a pump-action water gun. Within four weeks he has raised enough money. Wayne is excited that all his hard work will allow him to buy the gun and have lots of fun. His parents are pleased because, for the first time, Wayne has helped keep the house tidy. The toy shop owner is pleased because he will sell another gun. Maisie, however, is not sure. Wayne has been too busy to bother her for the last four weeks and that has been good, but now she is concerned how he will use the gun. She is worried about her pet cat getting wet and she is not sure that the manufacture of plastic toys is good for the environment.

This Winners and Losers Chart identifies some of the people directly and indirectly affected by Wayne's purchase of the pump-action water gun. Who do you think are winners and who are losers? Who else will be affected?

Performance testing

Evaluating a product will involve comparing how well it works against its performance specification. You have to ask, does it do what it was designed to do? Here is an example.

Doctor and dentist waiting rooms often have magazines and toys available. By the end of a busy surgery the waiting room is often a mess with toys and magazines left in a state of disorder.

The design specification for a storage unit for these items in a waiting room describes what it should do and provides a checklist against which we can assess its performance.

Here is the design specification for a waiting room storage unit.

> **What it has to do**
> It has to provide a means for storing magazines, newspapers and toys neatly.
>
> **What it should look like**
> It should reflect the decorative style of the room.
>
> **Other requirements**
> - easy access to items;
> - easily maintained;
> - manoeuvrable;
> - safe for all users;
> - encourage appropriate use.

The following questions will help you compare the performance of the unit against its specification.

- Does it accommodate all the magazines, newspapers and toys?
- Is the arrangement neat?
- Does it look good in the room?
- Is there easy access to the items?
- Is it easy to keep clean and tidy?
- Is it manoeuvrable?
- Is it being used?
- Have there been any accidents?

You can find the answers to these questions by watching people use the storage unit and by asking them questions about it. Don't forget to ask the people who clean the waiting room at the end of the day!

SRT6

Is it appropriate?

Appropriate technology is suitable technology. You can use these questions to find out if a product or technology is appropriate.

- Does it suit the needs of the people who use it?
- Does it use local materials?
- Does it use local means of production?
- Is it too expensive?
- Does it generate income?
- Does it increase self-reliance?
- Does it use renewable sources of energy?
- Is it culturally acceptable?
- Is it environmentally friendly?
- Is it controlled by users?

It is unlikely that any product or technology will score highly against all these questions. Many will seem appropriate in one context and inappropriate in another. Here is an example.

Injection moulding is a process which has revolutionized the plastics industry. Our lives are made easier by the production of cheap plastic goods which cater for our everyday needs. These include food containers, general household goods, toys and shoes. Millions of children have benefited from hours of constructive play using injection-moulded building bricks. Leather-soled shoes which once had to be made by hand and repaired regularly have been replaced by shoes with injection-moulded plastic soles which are cheap and hard-wearing.

This technology has created employment and wealth in many communities throughout the world where it clearly benefits both manufacturers and consumers.

However, in one community in North Africa the advantages of injection moulding have not been so obvious.

For this group of people the introduction of injection moulding to produce cheap plastic sandals has created unemployment and poverty as it has displaced the traditional leather sandal industry. The manufacture of leather sandals involved the whole community and relied on local sources of labour and natural materials. It supplied a local market which had a constant need for new leather sandals.

The introduction of the injection moulding process created a more attractive product for this market as the plastic sandals were cheaper and lasted longer. There was soon no local demand for leather sandals and the traditional industry closed. There was also less work as fewer people were needed to produce the plastic sandals and the synthetic materials used for manufacture had to be imported. More energy was needed to power the machines with a consequent increase in pollution. The new plant was owned and run by a foreign company so little of the profits created by the industry were shared with the local community.

It is clear that for many people in this community injection moulding was an inappropriate technology.

Whether a product or technology is appropriate will depend upon the situation in which it is used.

Strategies

Modelling Materials Chooser Chart

Here are some suggestions to help you choose which modelling material to use for modelling design ideas for each line of interest.

Line of interest	Modelling techniques	Model
Seating	Wooden frames	Wooden frame chair
Storage	Block foam model	Foam model of desk tidy
Lighting	Paper shapes and wire frames	Wire frame lamp with shade
Toys and games	Card shapes with found materials	Card model of activity centre with buttons and sweets for control features
Automata	Construction kits	Simple automata mechanism
Body Adornment	Found materials	Necklace made from found items
Testing	Rapid prototyping	Test-bed structure testing rig

StRT1   StRT3   StRT3

Strategies Chooser Chart

This Chooser Chart gives you information about strategies:

- when to use a strategy in a Capability Task;
- how long the strategy will take;
- how complex it is;
- whether it involves other people.

Use the key to find out what the icons mean.

Key to icons:

When: beginning – middle – end

Time: short – long

Complexity: simple to complex

Other people: one other to many

Strategy	Comments
Identifying needs and likes	
PIES	beginning · short · simple
observing people	beginning · short · simple
asking questions	beginning · short · simple
using books and magazines	beginning · short · simple
image boards	beginning · medium · complex
questionnaires	beginning · long · complex · many
Design briefs	beginning · short · simple
Specifications	beginning · short · complex
Generating design ideas	
brainstorming	beginning / middle · medium · simple · one other
attribute analysis	beginning · short · simple
observational drawing	beginning / middle · medium · simple
investigative drawing	beginning / middle · medium · complex
Modelling	
modelling appearance	middle · medium · complex
modelling performance	middle · medium · complex
modelling with computers	middle · long · complex
Applying science	middle · medium · complex
Systems thinking	middle · medium · complex
Planning	middle / end · short · simple
Evaluating	
user trip	end / middle · short · simple · one other
winners and losers	end / middle · short · simple · one other
performance testing	end / middle · short · complex · one other
appropriateness	end / middle · short · simple · one other

In the world of business and industry design proposals can only be turned into saleable products if the designers communicate their proposals effectively. Designers have to communicate their ideas to clients and manufacturers as shown below.

▶ *Presenting and discussing design proposals*

The designer

I work for a large furniture manufacturer. I am part of the in-house team of designers that develops new products. It is the client who decides whether the design will be produced or not. So we need to present our ideas as effectively as possible.

The client

I am the manager of the furniture company. I and the rest of the management team are always looking to enlarge our product range. We employ a team of in-house designers. We have to be sure that the new product ideas the team presents to us will be what people will buy.

The manufacturer

When a new product is proposed I have to be involved at an early stage in planning the manufacturing process. I need to be sure that the factory can make the product quickly and efficiently to the required quality standard.

In your work you will be both the designer and the manufacturer. You may even be the client as well. Do not fall into the trap of thinking you don't need to communicate your ideas just because you know what you are doing! You have to communicate your ideas so that they can be understood by other people. Otherwise you may find that you have not worked out all the details needed to turn your design idea into a real product that works well, looks good and is easy to use. You can use the techniques in this unit to help you communicate your ideas.

Rendering

To give a more realistic and eye-catching quality to drawings a designer will often **render** them. This term applies to any technique which makes the drawing look 3D and shows surface textures. You can use a wide range of media to do this.

Metal and plastic

Line shading

Using a black pen or pencil draw a series of parallel lines on the surfaces of your drawing. The spaces between the lines can be increased or decreased to give an impression of light or dark. It is usual to make upward facing surfaces lightest.

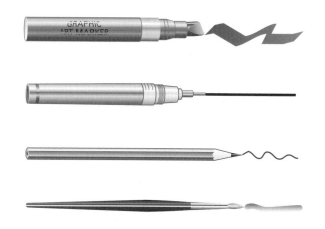

> ◗ *This technique suggests the surface of materials such as metal or acrylic which are smooth and even in texture*

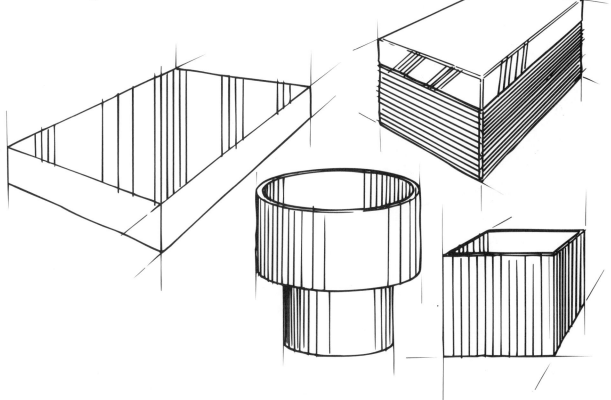

> ◗ *Line shading is also effective for indicating the form of an object. The shading shows that these objects are hollow*

> ◗ *To show that a surface is shiny or reflective, use vertical lines rather than lines which follow the edges of the object*

Communicating your design proposals

Wood textures

The drawing of the toy car shows the different grain patterns caused by the way the wood has been cut from the original piece. End and side grains tend to look closer spaced than the more open and attractive facing grain.

Coloured pencils have been used to do this drawing. A background colour has been lightly applied before the darker grain pattern has been added. Compare this with the effect created with pencil and pen.

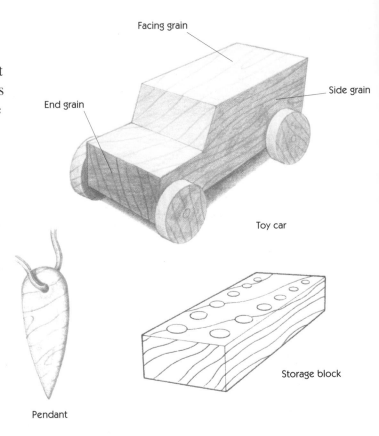

Facing grain

Side grain

End grain

Toy car

Storage block

Pendant

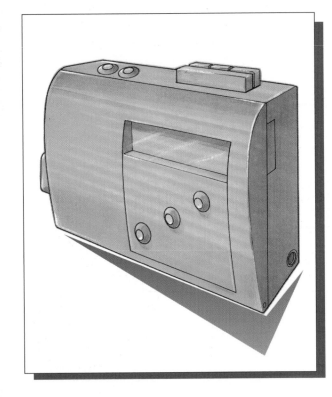

Felt-tip marker rendering

To use markers well you will need to practise before working on a special piece of work. Here are some tips which will help you.

- Work quickly and evenly – don't rest the pen on the paper.
- Don't worry about going over the edges of your drawing – see next point.
- Cut out your finished drawing and remount it to obtain clear outlines.
- Use coloured pencils to add details and white pencil for reflections.
- Use white paint for highlights.
- Use only a limited range of colours or a variety of shades of the same colour.

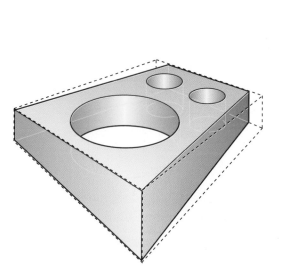

Crating

This technique helps you to draw objects which look complicated at first sight. You have to imagine that the object is packed tightly in a box or crate. First you draw the simple box shape in the correct position and proportions. Then you modify this and add details so that the shape 'inside' the box becomes the object.

You can apply this technique to any size of object although sometimes it is necessary to break the object down into several box shapes or cubes.

The chair shown here looks rather complicated. If, however, it is visualized as two separate boxes on top of one another with another box leaning at an angle behind it, it becomes a lot easier to draw.

Notice that the technique can be applied to most drawing systems. The chair is drawn as an isometric projection whereas the jewellery below uses different perspective views.

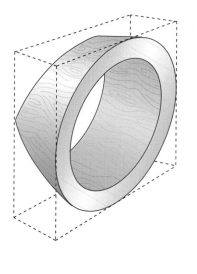

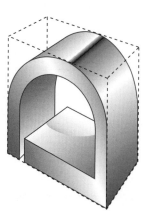

▶ *Designs for jewellery using crating*

Communicating your design proposals

Shadows

Shadows give us visual clues about the shapes and forms we are looking at. When a designer is presenting ideas to a client the addition of cast shadows makes the drawings more understandable and also more realistic.

There are three factors which give the shape and position of a cast shadow.

1 The position of the light source.

2 The shape of the object.

3 The form of the surface on which the shadow falls.

Shadows provide a lot of information

Constructing cast shadows

You can use steps 1–5 to help you construct cast shadows.

1 Draw the box.

2 Put in the light source and the shadow vanishing point. This will be a point vertically below the light source on the surface on which the object is standing.

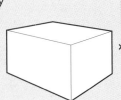

3 Draw lines from the light source through the corners which will cast shadows.

4 Draw lines from the shadow vanishing point through points on the ground vertically below the corners which cast the shadows.

5 Shade in shadow area.

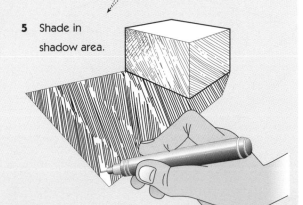

Perspective drawing

Perspective drawing gives the most realistic view of an object. This is because it takes into account the fact that when you view an object, certain lines appear to converge at a vanishing point.

The three most common forms of perspective drawing are shown here.

- **Single-point** perspective is useful when you want to draw an interior.

- **Two-point** perspective is often the most useful method because it can give a clear view of three sides of an object.

- **Three-point** perspective gives an aerial view and an impression of architectural scale.

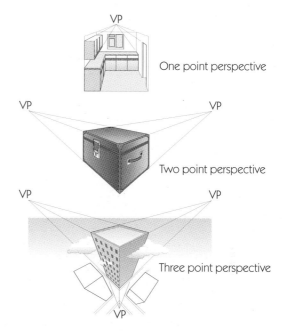

One point perspective

Two point perspective

Three point perspective

Two-point perspective step by step

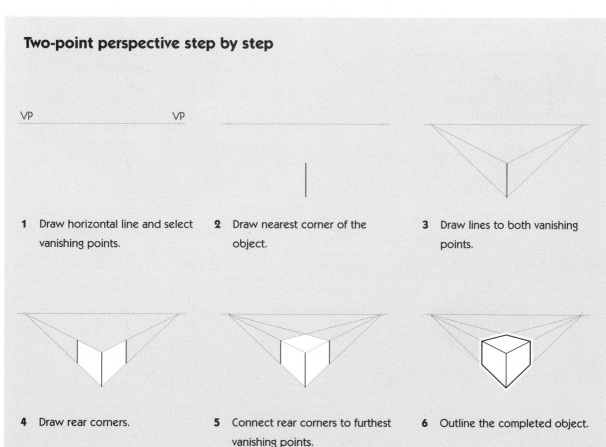

1 Draw horizontal line and select vanishing points.

2 Draw nearest corner of the object.

3 Draw lines to both vanishing points.

4 Draw rear corners.

5 Connect rear corners to furthest vanishing points.

6 Outline the completed object.

You can produce an isometric drawing by following these steps using a 60/30 set square.

1 Draw a baseline and construct the nearest corner of the object at 30° to the baseline.

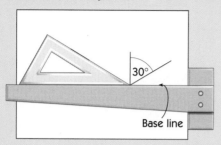

30°

Base line

2 Construct the crate by drawing lines parallel to the three corner lines.

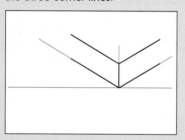

3 Use a ruler to mark off the correct measurements to construct your drawing.

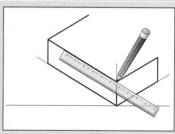

4 Add final detail.

Isometric drawing

Isometric means 'equal measure'. In an isometric drawing lines of equal length along the axes in the object being drawn appear as such in the drawing. This method of drawing is suitable for computer-aided design (CAD) where drawings are entered into the computer as a series of numbers or measurements.

Costing materials and components

It is important to show the cost of materials and components needed for your design proposals. You can use a costing chart like the one below to do this.

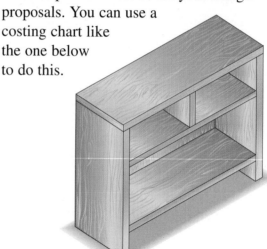

	Material	Dimensions	Pieces off	Approx price	Cost
1	Veneered chipboard	870 x 600	1	£10 per m²	5.20
2	Veneered chipboard	900 x 200	1	£10 per m²	1.80
3	Veneered chipboard	870 x 185	2	£10 per m²	3.20
4	Veneered chipboard	600 x 200	2	£10 per m²	2.40
5	Veneered chipboard	270 x 185	1	£10 per m²	.50
6	S/A blocks		18	8p each	1.44
7	Cross-headed chipboard screws	25 x 8	60	£4 per 100	2.50
8	Iron-on wood veneer edging	4200 mm	1	48p per metre	2.00
9	Varnish		1	£4.20 per 250 ml	4.20
				TOTAL COST	£23.24

Orthographic projection

You should use the drawing system called orthographic projection for accurate scale drawings of your design ideas. These drawings are called **working drawings** and are based on square-on views of the object. As you can see you can obtain six square-on views when you look at a camera.

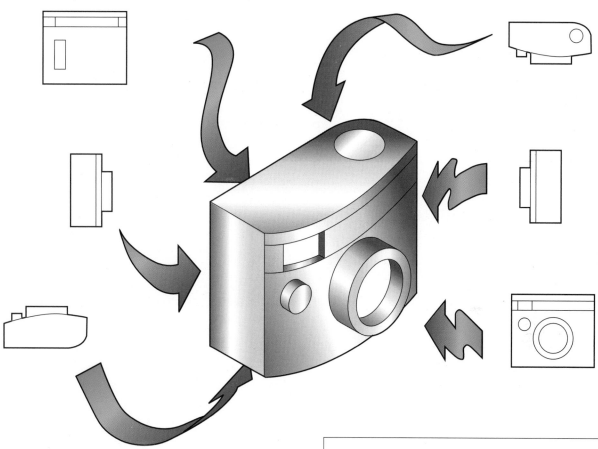

Usually you only need to draw three views to give enough detail about your design. There are two ways of arranging these views. They are called **first-angle projection** and **third-angle projection**. Each has its own symbol. You must always show whether your plans are first- or third-angle projections.

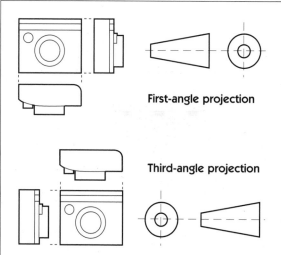

First-angle projection

Third-angle projection

Drawing a first-angle projection

You will need to use a drawing board with either a T-square or parallel motion for drawing parallel lines. You will need set squares, compasses and dividers as well as a sharp pencil.

Here are the step-by-step instructions for drawing a first-angle orthographic projection of a bracket used to hold a motor in an automaton.

1 Draw the front elevation on a baseline.

2 Draw in vertical and horizontal projection lines from important features.

3 Draw in the plan view.

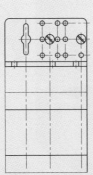

4 Draw in the horizontal projection lines from important features and a line at 45° from the front elevation across the horizontal projection lines from the plan view.

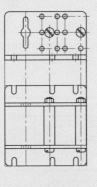

5 Draw in vertical projection lines from the points where the 45° line cuts the horizontal projection lines.

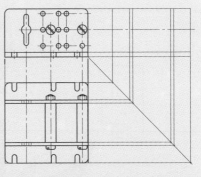

6 Use the crossing points of these vertical projection lines and the horizontal projection lines from the front elevation to construct the end elevation.

7 Add labels.

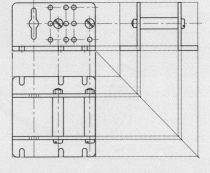

A MOTOR BRACKET

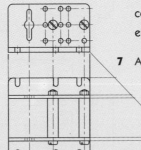

FRONT ELEVATION   END ELEVATION

PLAN

Using British Standards conventions

In industry it is unlikely that the person making a product is the same person who drew the designs so it is important that the designer produces drawings which communicate his or her ideas clearly. These kinds of drawings are called **working drawings** and should be set out in a clear and organized manner. The British Standards Institution (BSI) gives a set of rules (conventions) for such drawings.

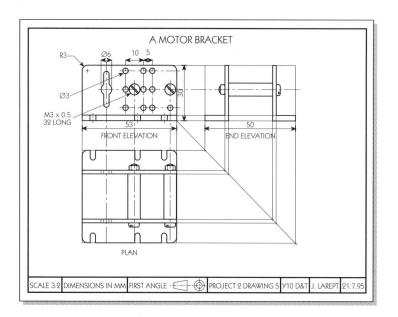

Some common abbrebiations	
Ø	Diameter
R	Radius
mm	Millimetre
cm	Centimetre
m	Metre
CSK	Countersunk
O/D	Outside diameter
I/D	Insider diameter
RDHD	Roundhead
DRG	Drawing
Matl	Material

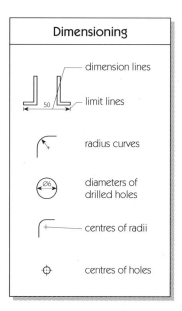

Dimensioning

- dimension lines
- limit lines
- radius curves
- diameters of drilled holes
- centres of radii
- centres of holes

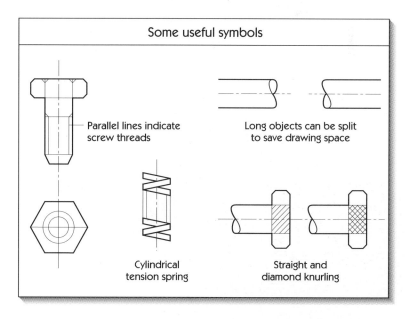

Some useful symbols

Parallel lines indicate screw threads

Long objects can be split to save drawing space

Cylindrical tension spring

Straight and diamond knurling

⑤

Sectional views

It is sometimes important for drawings to show the insides of products. You can use a sectional view to do this. You make an imaginary cut through the object, remove one half and draw what you can then see.

The 'cut' surfaces are usually hatched with regular lines at approximately 45°. These lines change direction between different pieces of material.

◗ *Sectioning to reveal interiors*

Section plane indicated by **A — · — A**

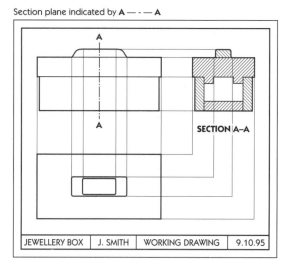

A

A

SECTION A–A

| JEWELLERY BOX | J. SMITH | WORKING DRAWING | 9.10.95 |

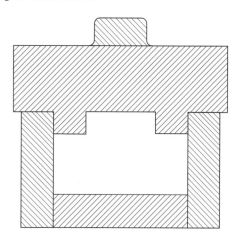

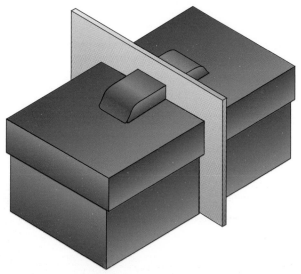

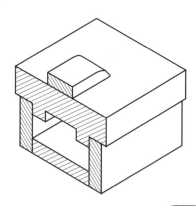

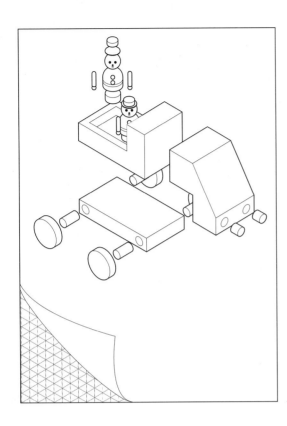

Assembly drawings

You can use an assembly drawing to show how the different parts of a product fit together. This type of illustration is sometimes called an **exploded view**. It is particularly useful for products with lots of working parts like a watch.

Hints for providing successful assembly drawings.

- Use isometric grid paper for guide lines.
- Keep parts on the same axis (in line with one another) when exploded.
- As far as possible show each piece separately.
- Avoid overlaps.
- Keep all parts exploded along an axis in their correct relative positions.
- Sketch out a plan of your drawing before committing yourself to any detailed work.
(It is easy to run out of space with this type of drawing.)

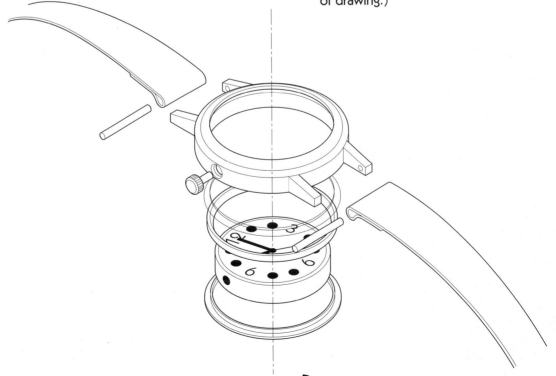

This exploded view shows how all the parts fit together

Which sort of drawing?

One system of drawing will rarely communicate all the information a designer wishes to convey to the maker or client. You will usually need to use a combination of systems. Even then there will be times when additional information is necessary. Sometimes you will need to add written notes specifying a particular process or finish such as sand-blasting or dip-coating. You may need to add extra drawings. Here are some examples taken from drawings of a Walkman.

You can use this Chooser Chart to help you decide which technique or drawing system to use.

What you want to communicate	Techniques or drawing systems to use
surface appearance	rendering on drawings
overall appearance/proportion	crating
extra realism	cast shadows
an interior	single-point perspective
realistic appearance	two-point perspective
aerial view	three-point perspective
scale drawing suitable for CAD	isometric drawing
cost of materials and components	costing chart
details for making	orthographic projection
internal details	sectional views
how parts fit together	exploded views
special features	hidden detail, animation for moving parts, enlarged detail

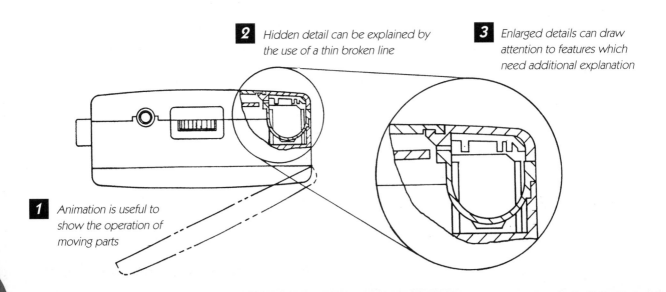

2 Hidden detail can be explained by the use of a thin broken line

3 Enlarged details can draw attention to features which need additional explanation

1 Animation is useful to show the operation of moving parts

Seating

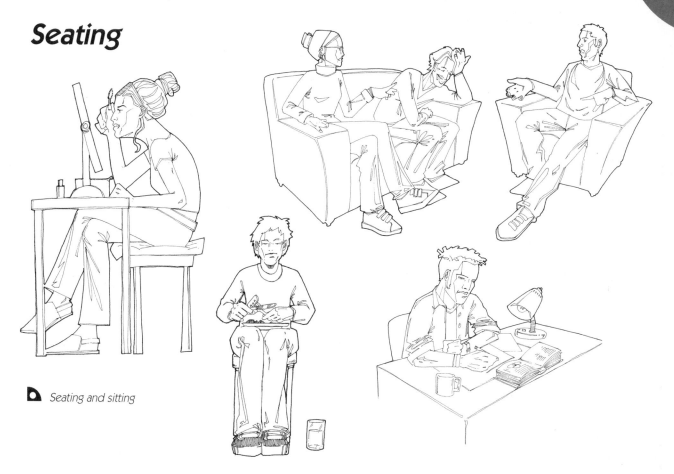

▶ *Seating and sitting*

Just sitting down?

We spend a lot of our time sitting down. Here is a list of just some of the things we do that involve sitting down:

- relax and chat with friends and family;
- watch TV, listen to music or the radio;
- eat 'off our laps';
- eat at a table;
- hold a meeting around a table;
- read and write at a desk;
- use a computer at a desk;
- do practical activities at a worktable;
- put on make-up at a dressing table.

We sit down in all sorts of places:

- at home;
- at school;
- at work;
- in cinemas and theatres;
- in cafes, restaurants, bars and pubs;
- in cars, buses, trains and aeroplanes;
- in the garden;
- in parks or recreation grounds;
- at picnics;
- on the beach.

The type of seating you use depends on where you are and what you are doing. When you design seating you will therefore need to take into account where it will be used and what the people using it will be doing.

Designing for comfort

If a chair is the wrong size or shape it will be uncomfortable. You can use British Standards (BS) information to find out sizes and shapes to use in your designs. Information about an office chair is shown below.

You will need to have similar information to this for any chair you are designing. You should make sure that the information you have is about the sort of chair you want to design and for the type of person you are designing for. An easy chair for an elderly person will have a different shape and size from a school chair for a seven year old.

It is sometimes important to use padding on armrests, backrests and seats. The padding will reduce jarring on sitting down and provide a softer surface. You can use polyurethane foam for this, as shown below. You will need to make sure that it is thick enough to provide sufficient padding without being so thick that it makes getting up difficult.

You will need to cover the padding in upholstery fabric. This protects the padding from wear and adds to the appearance of the seat.

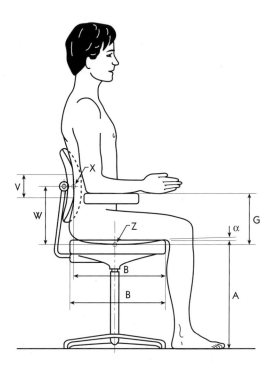

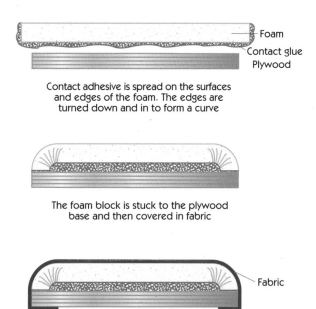

▶ Adding upholstery to a seat

Contact adhesive is spread on the surfaces and edges of the foam. The edges are turned down and in to form a curve

The foam block is stuck to the plywood base and then covered in fabric

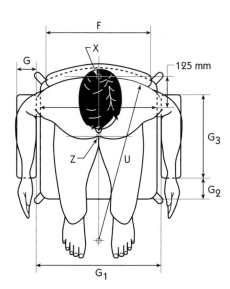

▶ Important dimensions in an office chair (from BS5940)

Designing for strength and stiffness

Whatever sort of seat you design it will need to be strong (so that it does not break when people sit on it) and stiff (so that it does not deform too much when people sit on it). You can imagine your seat as being made from a combination of different structural elements:

- flat surfaces or platforms (slabs)
- frameworks
- shell forms.

You will need to think about ways in which these structural elements can be combined to give a design which is both stiff and strong. To help you, the panel shows several different seating designs in terms of these structural elements.

Once you have worked out a design in these 'broad sweep' terms you will need to think about the shapes and sizes of the parts making up the framework, the way these are joined together and the way shell forms and flat surfaces are joined to the framework. The section on structural systems (page 171–179) has information to help you with this. You will also need to think about which materials to use. The Chooser Charts on pages 188–194 will help you here.

What about mistreatment?

Often people use seating in ways for which it was not designed:

- they stand on chairs;
- they lean back in them and tip them over;
- they use them to prop doors open;
- they use them to hold up planks for decorating;
- they play party games with them.

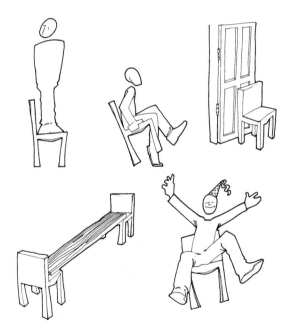

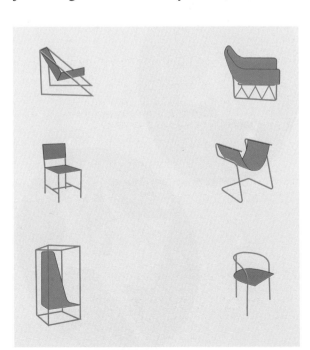

Well designed seating will be able to withstand most of this ill-treatment. Think carefully about what might happen to the seating you design and make sure it is not too flimsy.

Design guides – seating

Designing for stability

A person risks serious injury if they fall off a chair or a chair tips over, so it is important that any seating you design is stable. You can find out more about designing for stability in the section on structures (page 172).

Folding seats

It is often useful for a seat to fold away. Deckchairs used in parks and on beaches, garden chairs, examination chairs, camping chairs all fold away. These chairs are mechanisms as well as structures. If you design a fold-away chair you will have to think about each of the following:

- how the chair opens up;
- how it is held rigid;
- how it can be released;
- how the chair folds down.

You will find that you can work out most of these details by making simple working models of your design ideas. The models shown here are made from card, balsawood, masking tape and pins. They took only 40 minutes to make. They showed the designer that her idea would work and helped her see the sort of joint she would need for each of the moving parts.

Designing for durability

If you choose a covering for your seating you will need to make sure that it is hard-wearing as well as attractive. It is important that it does not wear away or tear easily. You can do simple rub tests to compare different fabrics. The care with which you work and finish the materials will also play a large part in the overall durability of your seating.

▷ *These models showed the design idea was a good one*

112

Thirty years of chairs

The panel below shows thirty years of chair design. You can use these designs as a source of inspiration for your own work. To help you understand them sketch each one in terms of flat surfaces or platforms, frameworks and shell forms and ask yourself these questions.

- Who is likely to have used the chair?
- Where is it likely to have been used?

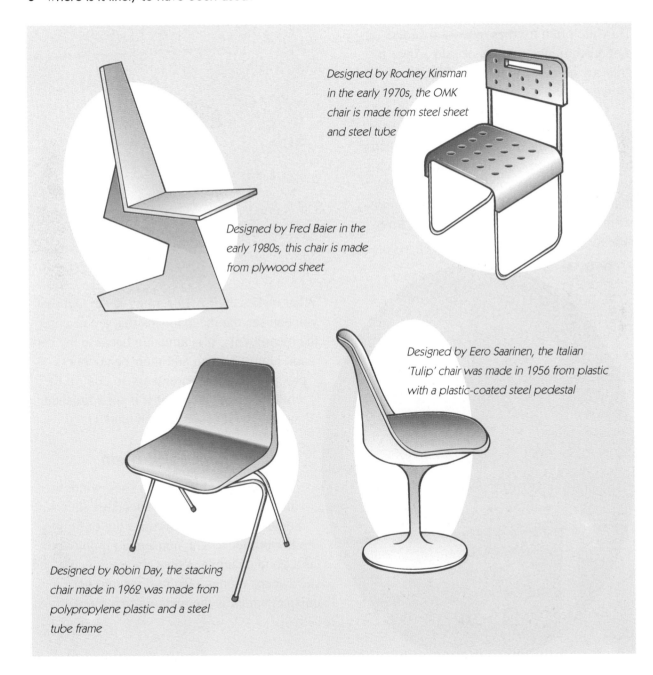

Designed by Rodney Kinsman in the early 1970s, the OMK chair is made from steel sheet and steel tube

Designed by Fred Baier in the early 1980s, this chair is made from plywood sheet

Designed by Eero Saarinen, the Italian 'Tulip' chair was made in 1956 from plastic with a plastic-coated steel pedestal

Designed by Robin Day, the stacking chair made in 1962 was made from polypropylene plastic and a steel tube frame

Automata

What are automata?

Automata are mechanical toys that perform a series of movements automatically. Once set in motion they go through a routine which is usually designed to be intriguing or amusing. They are often used as 'attention getters' at demonstrations or points of sale. There is a growing market for automata as 'toys for grown-ups'.

The two cats automaton shown below is a typical example. It is operated by an electric motor. The large cat turns a handle which seems to cause the small cat to move. The legs, arm and head of the small cat all move while the eyes of the large cat move from side to side.

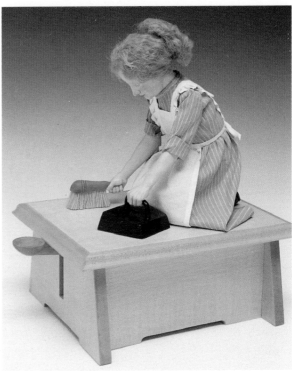

What might the mechanism be?

What makes the toy intriguing is the fact that you can see the mechanisms that are causing the movements. It is amusing because the cats look almost human and their behaviour parodies the way parents try to control their children. Note also that it is well finished: no rough edges or scratch marks.

Where I might see them

You can find automata in shop window displays, museums and specialist shops. Sometimes simple automata are built into other products. The housemaid money-box is a good example. When the lever is pressed she sweeps money off the floor into her dustpan which is the opening to a money-box.

Can you see how the mechanisms make the cats move?

Who is it for?

It is important that you are clear about who an automaton is for. It could be for a particular group of people: young children, older children, teenagers or adults. Or it could be to attract attention to something – an event, a product or a service.

If the automaton is for a particular group of people then you will need to think about what might appeal to them. There are some examples below. Can you work out who each might be suitable for? If it is to promote something then you will need to make sure that its theme is in keeping with this.

◗ *Who would like these automata?*

Developing the idea

It is important that the movements of the
automaton are appropriate. For example if
the figure is an animal then the movements
should be based on the way that animal moves
and behaves in real life. So for a dog the tail
should wag rapidly and the mouth open and
close as if it were barking. For a cat the tail
should move slowly from side to side and the
mouth should open and close more slowly to
imitate meowing rather than barking.

It is useful to start by carefully observing
animals and people so that you can describe
the movements you want. Of course you can
make your automata funny by deliberately
choosing movements that are inappropriate.
This often works well, for example if you
make animals behave like humans.

Designing the mechanism

Once you have an idea of the scene and the
movements you want you will need to think
about the mechanisms you need to achieve
them. You can use these questions to help you
describe the mechanisms you need.

	Input	Output
● What type of motion?		
● What direction of motion?		
● What axis of rotation?		
● What range of motion?		
● What speed of motion?		
● What sizes of force are involved?		

Once you have a clear picture of what the
mechanisms have to do then you can use the
Mechanisms Chooser Chart to identify the
different mechanisms you need. When you
think you have a good idea of these, use the
modelling techniques on pages 81 and 83 to
work out the details.

Supporting the mechanisms and the figure

Your automaton will need to stand on some sort of frame or box. This will serve two purposes.

First, it will house the mechanism so you will need to think carefully about its size. You will also need to consider where any axles and springs are held in position and where any push rods and wires pass through the base onto the automaton. Accurate alignment is very important and you will need to plan your making carefully to achieve this. You may find it worth making your mechanism adjustable so that once you have set it up you can alter it slightly to get the very best performance.

Second, the support can reveal or hide the mechanism. You have to decide whether to keep the mechanism hidden or to share its secrets with the user.

Remember that the support is part of the automaton and will be seen by the people who use it. Make sure that it is well finished and attractive.

Making it work

Many modern automata are operated by hand. In this way the user can control the speed of operation and become involved with the machine. Clockwork motors allow the user to be involved (winding up the motor) and to step back and watch what happens. You can use electric motors powered by batteries and control the motors with switches.

You can also use electricity to produce light and sound, e.g. flashing eyes and animal roars! One or two designers have used falling water or rising hot air to turn turbines as a source of power.

You can use these questions to help you work out which method is best for your automaton.

- How will you make sure that the method you use is powerful enough?
- How will you make sure that it does not get broken easily?
- What would the user like to use to make the automaton work – a lever to pull, a handle to turn, a switch to turn, push or click, a key to wind up?
- How will the user know what to do?

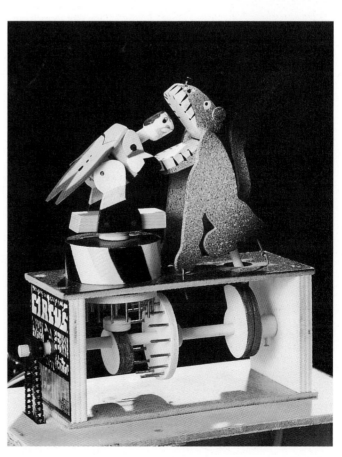

Lighting

What is the purpose of your light?

To understand the purpose of a light you need to answer the following questions.

- Who is the light for?
- Where is it going to go?
- What is its function?

The answers to these questions will give you a clear design brief.

What are the different types of light?

You will need to choose the type of lighting unit which will meet your requirements. One way to decide on the type of light you need is to imagine different sorts of light in the place where the light is to go. The most appropriate type of light should then become obvious. You can use the descriptions in the panel to help you do this. It is possible to get lights that are combinations of different types.

A **floodlight** is designed to flood an area with light. The light spreads out at a wide angle to illuminate an area.

A **downlighter** is the most common type of light. It consists of a lighting source which is fixed in a high part of the area being lit. It shines down and illuminates things beneath it.

A **spotlight** is designed to concentrate a beam of light at a narrow angle. It is used to illuminate a specific place or object.

An **uplighter** is used where an area requires more subtle and less direct lighting. The light is low down and shines up.

Task-lighting is designed to provide illumination for someone carrying out a particular activity, and to reduce eye strain. This type of lighting is usually directed onto the activity.

Choosing a light source

In the interests of safety you will be restricted to a low-voltage direct current (d.c.) supply for your design. This still gives you a wide range of light sources to consider. Some of these are described in the table below.

The table shows that you can get a wide range of lighting levels by choosing different light sources or by combining light sources. You can investigate the amount of light you need and find a light source to provide this.

Remember:

- Light sources have recommended voltages. If you exceed the recommended voltage you will damage or destroy the light source.

- Light sources draw a particular size of current when operating at the recommended voltage. The larger the operating current the more quickly will the light source drain a battery.

- The greater the power of the light source (the greater the number of watts) the brighter it shines. A typical domestic light bulb has a power of 60 W.

Light source	Voltage	Current	Power	Dimensions and fittings	Cost
Filament lamp					
	6.5 V d.c.	300 mA	2 W	MES, tubular or round lamps length 30 mm max diameter 12 mm max	low
	12 V d.c.	183 mA	2 W	MES tubular or round lamps length 30 mm max diameter 12 mm max	low
	12 V d.c.	4.0 A	48 W	SBC, candle or globe lamps length 80 mm max width 30 mm max	medium
Tungsten halogen lamp					
	6 V d.c.	1.0 A	6 W	flanged length 31 mm max diameter 9.3 mm max	medium
	4 V d.c.	0.5 A	2 W	MES length 31 mm max diameter 9.3 mm max	medium
	5.2 V d.c.	0.85 A	4.4 W	MES length 31 mm max diameter 9.3 mm max	medium
	12 V d.c.	1.66 A	20 W	2 pin open/closed face dischroic length 50 mm max diameter 100 mm max	medium

Key: MES - miniature Edison screw; SBC - small bayonet cap.

You can find more information about light sources from both the RS Components and the Maplins catalogues.

Stability, adjustability and elegance

You will need to ensure that your design meets the following criteria.

- **Stability** If free-standing it should not topple over easily. You can find out more about designing for stability in the section on structures (page 172).

- **Adjustability** If the light can be adjusted then it is important that it remains stable in all possible positions. You can find ideas for making moving parts in *Ways to make your product* (page 196) and *Mechanical systems* (page 141).

- **Elegance** It is important that your light looks attractive in its setting and appeals to the user. You can achieve this partly in the overall design, paying particular attention to the organization of the electrical wiring and the choice and position of any control switches. The care with which you work and finish the materials will also affect the final appearance of your light.

Supporting the light source

The light source will be held in a light fitting. To support the light source you will need to work out a way of holding the fitting. The holder could take a single fitting or a cluster of fittings. You will need to work out how the holder is joined into the rest of the design and how the wiring travels from the holder to the power source and to any switches.

Directing the light

You can use a shade or a reflector to direct the light. You can make shades from paper and wire or thin thermoplastic sheet. These materials are effective because they are translucent and appear to glow when the light is on in a darkened room.

Take great care to ensure that these materials will not get hot in your design as they are both inflammable. You can make shades from thin sheet metal which is not inflammable, but it is not opaque and will direct quite a harsh light. You can polish sheet metal so that it acts as a reflector, or use metallized thermoplastic sheet, but you must ensure that it does not get hot.

Switches

You will need to think about the position, appearance and type of switch you choose to ensure that your light works, looks good and is easy to use.

You can use fittings like this to hold the light source

Choosing a power source

You have three choices for the power source:

- non-rechargeable battery;
- rechargeable battery;
- low-voltage supply unit powered by mains electricity.

Non rechargeable batteries

These are the least expensive option in the short term. To provide lighting for any length of time you will need to use alkaline batteries. These are available in a range of sizes – the larger the battery the longer it will last.

Non-rechargeable batteries are available as multiples of 1.5 V, from 1.5 V to 9 V terminal voltages. You can use battery holders for connection and to match the voltage of your light source. PP3 batteries can be connected using press-stud connectors.

Rechargeable batteries

These are more expensive and you will also need a charger unit, but these are once-only purchases. Rechargeable nickel-cadmium cells are available in the same sizes and terminal voltages as non-rechargeable batteries.

Low-voltage supply

Using a low-voltage supply powered by mains electricity is an expensive option. Also, you will need to make sure that it is checked by a qualified electrician.

In some situations there is a low-voltage supply readily available. In a car, for instance, the cigarette lighter socket can be used to connect to the car battery which has a 12 V terminal voltage.

Thinking about the electric circuit

The simplest circuit for a light consists of a light source, a battery, a switch and connecting wires. As soon as you put an extra light source in the circuit you will need to make decisions.

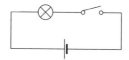

If you put the light sources in series then the switch will operate them both simultaneously but they will not shine as brightly as the single light source.

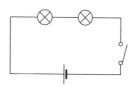

If you put the light sources in parallel then they will shine as brightly as the single light source but use the battery up twice as fast. You can add an extra switch so that you can control the lights independently.

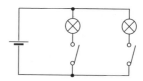

Or you can use a double pole/double throw changeover switch with centre-off position to allow you to have no, one or two lights on.

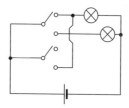

Remember:

- simple series for one bulb;
- parallel for more than one bulb;
- position of switches is important.

Storage

Why store?

People store things for one or more of the
following reasons:

- to keep things safe;
- to prevent losing things;
- to organize things;
- to allow things to be ready for use;
- to enable things to be checked.

When designing storage, it is important that
you identify which of these criteria are to be
met.

Designing for ease of use

Storage systems should be easy to use. The
information in the panel above shows where
items should be stored according to their
weight and frequency of use.

The size of items to be stored is also important
and you will need to measure things carefully
and make allowances for variations in size and
'clearance'. The panel below shows the sorts
of measurements you may need to make.

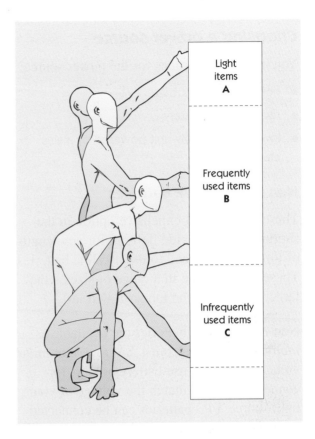

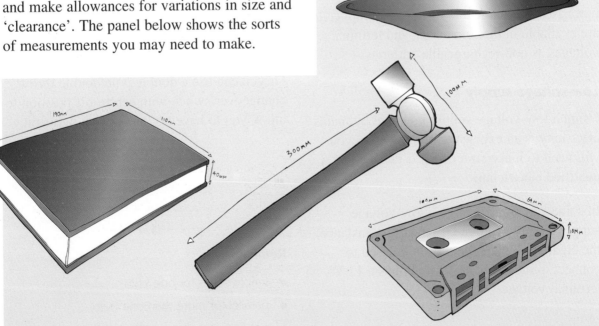

Storage opportunities in the home

The panel shows places in the home where storage will be needed, and activities that use items requiring storage. By looking at these places and investigating the activities you can find opportunities for storage.

Places

Bedroom
clean clothes, make-up, jewellery, dirty clothes, bedside books and magazines

Bathroom
first aid, medicines, dirty clothes, towels, washing materials, cleaning materials

Kitchen
crockery, cutlery, cooking utensils, cleaning materials, food and drink

Study
books, stationery, files, computer (keyboard, screen, printer), pens, pencils, rubbers, rulers

Dining room
crockery, cutlery, glasses, condiments, table linen, candles and candlesticks, alcoholic drinks

Activities

Washing clothes
dirty clothes, washing materials, iron, ironing board, clean clothes, pressed clothes

Gardening
bulbs, seeds, tools, seed trays, flower pots, fertilizer, compost, labels, ties, string, canes, stakes, hose, watering can

Maintenance
hand tools, electrical tools, portable work-bench, fixings and fittings, adhesives and fillers, abrasive papers

Decorating
paints, papers, brushes, rollers, trays, cleaners, pasting table, preparation tools and materials, electrical tools, step-ladder

Cleaning
vacuum cleaner, dustpan and brush, broom, bucket, mop, cleaning materials

Entertainment
TV, radio, video recorder, CD player, record player, tape player, vinyl records, compact discs, audiotapes, videotapes

Hobbies
tools, materials, reference books, portable workspace, lighting

Methods of storage

The panel below shows a range of storage methods.

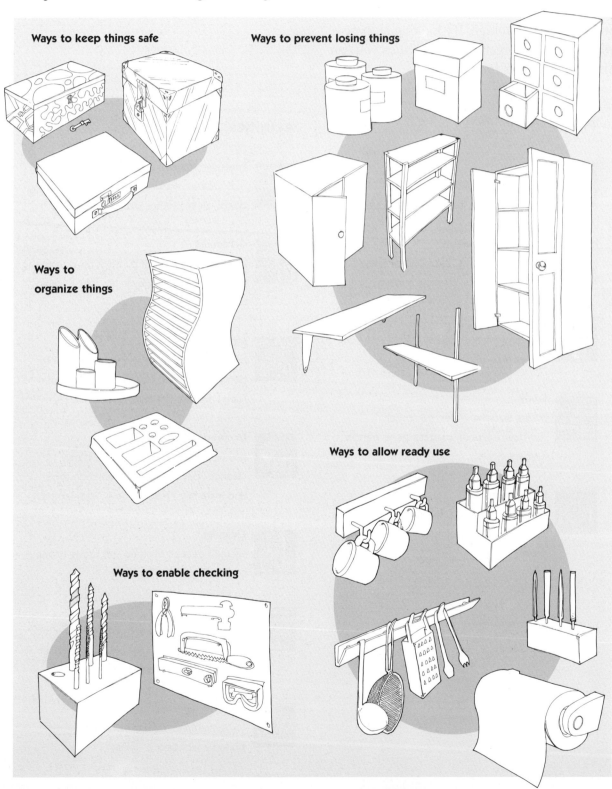

Ways to keep things safe

Ways to prevent losing things

Ways to organize things

Ways to allow ready use

Ways to enable checking

These questions will help you in your designing.

- **What will be stored?** By thinking about this you can decide the size of the storage system.

- **Where will it be kept?** By thinking about this you can work out what the storage system might look like and the sort of finish needed to protect it.

- **Who is likely to use it most?** If you know this you can find out what they will need from the storage system.

- **Who will also use it?** If you know this you can ensure your design is suitable for more than one person.

- **Does the design need to keep the stored items safe?** You may need to think about catches, locks and keys.

- **Does the design need to prevent the stored items from getting lost?** You may need to think about how to store the items near to where they will be used.

- **Does the design need to organize the stored items?** You may need to think about different ways to organize the items, for example according to shape, colour, size and use.

- **Does the design need to allow ready use?** You may need to think about how the stored items will be used.

- **Does the design need to enable checking?** You may need to think about a display system.

- **How much can I afford to spend on the storage system?** You will need to take into account materials, fittings and finishes.

Fittings

There are many components for storage systems that you should buy rather than make. These are usually called **fittings**. Some are shown in the table below.

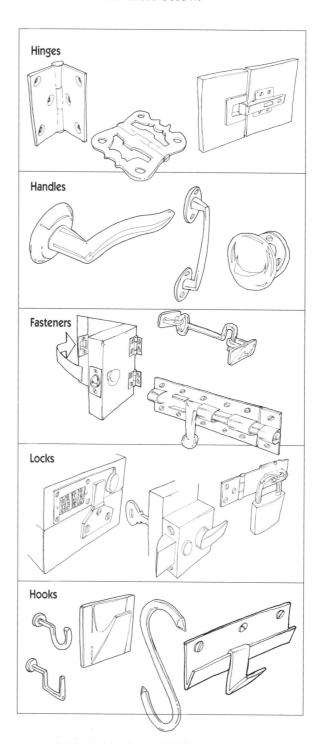

Toys and games

Playing a game

The picture shows a range of 'table-top' games. They are each based on competition, with the winner achieving a certain goal before the other players. Some of the games rely almost exclusively on luck; some depend very much on mental skill; others use a mix of both luck and skill. All the games shown fall into one of these three categories.

Designing a game

What makes a game attractive and appealing? These questions will help you decide why a game has appeal, and also to design a new game.

- **Who will play the game?**
  Age is particularly important. Is it for adults, children or families? This will affect the skills and knowledge used in the game. It will also help you identify a suitable theme for the game.

- **What is the purpose of the game?**
  to provide excitement;
  to provide intellectual stimulation;
  to test knowledge;
  to test skill;
  to inform or educate.

- **What combination of these makes people want to play the game?**
  This will help you design a game that people want to play.

- **What skills and knowledge will be needed or developed by the game?**
  You must ensure that the game is challenging but not daunting.

- **Is any luck required in the game?**
  If there is you will need to work out how it can be generated.

- **How do you play and what do you have to do to win?**
  This will help you work out the rules of play.

- **How will your game compare with existing games?**
  What will be different and what will be the same? What will make people want to play it?

Playing with a toy

The range of toys in the picture shows you that they vary widely. Some toys let you pretend to be someone or something else. Others represent someone or something and you have to work out what they do. Both of these types of play involve fantasy.

Sometimes toys involve learning a physical skill such as juggling, top spinning, kite flying, conjuring or diablo throwing. Some toys require physical activity like climbing, jumping, running, rolling, swinging, sliding or riding. Others involve building or constructing. Whatever the type of toy, it should be fun to play with!

Designing a toy

You can use these questions to help you design a toy.

- **What age range is the toy for?**
  This will affect the sorts of activity and behaviour used in playing with the toy.
  It will help identify the sort of toy that is appropriate.

- **What is the purpose of the toy?**
  to provide excitement;
  to develop physical strength and coordination;
  to provide intellectual stimulation;
  to develop knowledge and understanding;
  to develop social skills.
  What combination of these makes the toy fun to play with? This will help you design a toy that people want to play with.

- **What sorts of activities will be involved?**
  physical activity – pushing, pulling, flicking, poking, throwing, skimming, jumping, running;
  intellectual activity – reading, writing, drawing, puzzling, problem solving;
  social activity – talking, listening, questioning, watching, miming, acting, singing, dancing.
  The activities should match the purposes of the toy.
  This will help you decide whether the toy is for indoors, outdoors or both.
  It will also help you think about safety matters like safe landing surfaces, suitable climbing surfaces and sound structures.

- **What sorts of behaviour will be involved?**
  quiet and reflective;
  loud and rumbustious;
  fierce and aggressive;
  meek and passive;
  kind and caring;
  cruel and harsh.
  You will need to think about the sorts of behaviour that you want to encourage.

Get lucky

There are several ways of introducing luck into a game. Here are some possibilities:

- cards which are shuffled can be used to give bonuses, penalties, free turns and new opportunities;

- six-sided dice can be used to decide the number of moves or who goes first;

- spinners can be used instead of dice;

- spinning wheels can cause flashing lights and noise before settling on a number, bonus, penalty or prize.

Safety

It is important that toys and games are safe and do not harm those who play with them. You must ensure that any toy or game you design meets the British and European safety requirements. The *Compendium of British Standards for Design and Technology in Schools* contains useful information, as shown in the extract. You can obtain more detailed information from the British Standards Institution, Orders Department, Linford Wood, Milton Keynes MK14 6LE.

It is important that toys and games are not left lying around where they can get broken and trip people up. One solution to this problem is the toy box, which can be designed to be a toy in itself.

 Some useful information about toy safety from BS PP7302:1987

- Wood must be free from loose knots.

- There must be no sharp edges or splinters.

- Care is needed with hinges and folding mechanisms.

- Points and wires must be covered.

- Rigid parts that might stick out must be protected by firmly attached plastic or rubber caps.

- Non-detatchable parts must be fixed so that a child cannot grip them or so that they cannot be removed by a force of 90 N.

- Detatchable parts should not fit inside a choke monitor cylinder.

- Cords on pull-along toys should not have slip-knots or fastenings that can form slip-knots and should be at least 1.5 mm thick.

Toys and games for different ages

The table below describes the abilities and interests of children at different ages and the associated toys and games. You can use these as starting points for designing your own toys and games.

Abilities and interests of different groups	Associated toys and games
Infants (0–2 years)	
looking	soft books
grabbing/gripping	dangling/hanging shapes
feeling/cuddling	teddies
exploring/enquiring	rollers/rattles
seeing/discriminating	coloured balls/cubes
sucking/biting	teething toys
stacking	bricks
slotting	postboxes
hitting/banging	rubber/plastic shape-sorters
Early childhood (2–6 years)	
cuddling	teddies/dolls
looking/early reading	picture books
	pop-up/hidden books
constructing/changing/ modelling	jigsaws
	miniature environments
	simple construction kits, *e.g. Sticklebricks*
physical activity	bats and balls
	skipping ropes
	throwing/catching balls and rings
	hoops
social activity	simple board games, *e.g. Ludo*
	simple card games, *e.g. Snap*

Evaluating toys and games

The best way to evaluate a toy or game for a very young child is to watch the child using it – an observed user trip. Your observations can be in three parts.

- **First impressions**
  Does the child smile on seeing it?
  Does the child reach out for it?

- **Initial interest**
  How long does it take for the child to begin to play with it?
  Does the child need to be encouraged or helped to play with it?
  How long is it before the child loses interest?

- **Long-term interest**
  Does the child return to it during a playing session?
  How often? For how long?

You can use evaluations on existing toys or games as a starting point for your own design.

Later childhood (6–12 years)	
reading & writing	comics/story books plus associated figures
drawing & painting	
listening & talking/drama	masks
making & modelling	model kits
	complicated construction kits, *e.g. technical Lego*
collecting	small scale models, figures, facsimiles
keeping pets	
competition	adventure games
	more complicated board games, *e.g. chess, draughts, scrabble*
	more complicated card games
sports	bat/raquets and balls
	shooting games with targets

Testing equipment

Thinking about what you want to test

You will need to be clear about the following:

- why you are testing (reasons);
- what you are testing (objects);
- what you want to find out (properties or capacities).

Some possibilities for each of these are shown in the table opposite. Particular examples are shown in the illustrations.

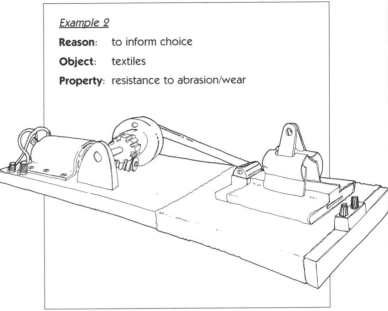

Example 2

Reason: to inform choice

Object: textiles

Property: resistance to abrasion/wear

D *John wanted to find out which fleece would be best for his mountaineering jacket*

Example 1

Reason: to evaluate fitness for purpose

Object: textile fastening in a product

Property: tensile strength

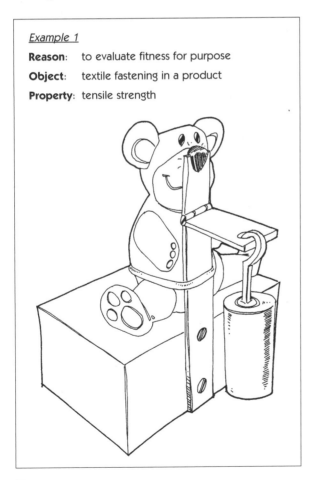

D *Sam wanted to find out the force needed to pull the nose off the teddy bear*

Example 3

Reason: to inform choice

Object: food materials

Property: knead-ability and effects of processes

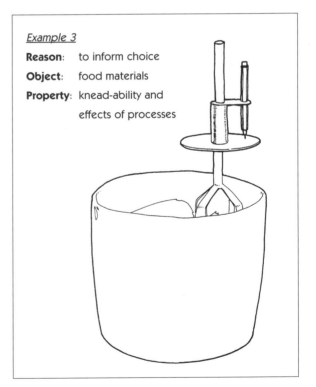

D *Michael wanted to see if certain additives made dough easier to knead*

LIRT7

Reasons	Objects	Properties	Capacities
to inform choice	construction materials	tensile strength	finger–thumb squeeze
to evaluate fitness for purpose	adhesives, joints and fastenings	compressive strength	finger–thumb pull
to find out what user's hands can do	textiles	torsional strength	finger–thumb twist
	textile fastenings	shear strength	hand squeeze
	biological materials, e.g. hair	fatigue strength	hand twist
	food materials	hardness	hand pull
	powders	toughness	
	liquids	resistance to wear and abrasion	
	packaging and protection	friction	
	products	stir-ability	
	human hands	knead-ability	
		flow-ability	
		pour-ability	
		squirt-ability	
		effects of processes	

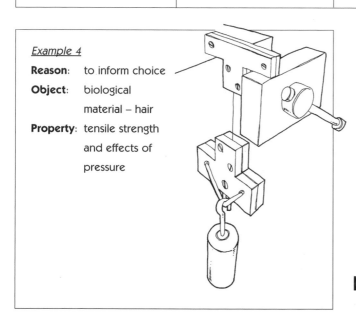

Example 4

Reason: to inform choice

Object: biological material – hair

Property: tensile strength and effects of pressure

▶ _Ruth wanted to know if highlighting made her hair more likely to break off_

Thinking about how to do the testing

You will need to think about each of the following points when you design your test equipment.

Destructive or non-destructive testing?

Destructive tests damage the object or sample so that it becomes unusable. After non-destructive tests the object or sample will still be usable.

What sorts of results?

Comparisons will tell you that one object or sample performs better than or differently from another object or sample. They will not necessarily give you a numerical record of how well each object performs. If you are comparing a number of objects or samples your results may be shown in a *rank order* as in the results of this hardness test:

Rank order of hardness is
1 – tool steel (hardened); 2 – tool steel (untreated);
3 – mild steel; 4 – brass; 5 – aluminium.

Numerical values will tell you more precisely how an object or sample performs. Here are some examples.

It took 5 minutes and 23 seconds for 50 g of sauce X (total sample size 500 g) to drip through a 1.5 mm diameter nozzle.

In trying to turn a 30 mm lid, person B was able to exert a torque (twisting force) of 0.5 Nm (newton metres).

Some tests give **approximate values**. For example:

Switch D was still working after 1500 on-off cycles but not after 2000 on-off cycles.

Some tests will allow you to **group** your results. For example:

Only 5 people in my class could exert a squeezing force of more than 25 newtons; 18 people could exert a squeezing force of between 15 and 24 newtons.

Descriptions can tell how an object or sample has been changed by a test. For example:

After 500 cycles of rubbing, fleece C was covered in balls of fibre and I could see light through it. In-between the balls it felt a lot thinner than an untested sample.

If tests are carried out to British, European or International Standards then other people know that they can rely on the results without needing to question the tester or check the apparatus.

How big and how many?

It will be important to choose a size of sample or object that can be managed conveniently. It is much easier to test the strength of steel using thin steel wire than by trying to measure the force needed to break a thick steel bar.

You will need to provide an accurate description of your sample or object so that other people can repeat your tests.

It will be important to carry out repeat tests so that you can check on the reliability of your testing method and the consistency of your samples or objects.

Getting a grip

How are you going to hold or contain your objects and samples? You may need to consider gripping methods (chucks for rods or threads; drums for threads; clamps for textiles, sheet and bar materials; Velcro for textiles; straps for irregular or fragile objects).

The nature of the test

What are you going to do to the sample or object? This will depend on what you are trying to find out. Here are a few possibilities:

- apply a constant force;
- apply a steadily increasing force;
- repeatedly apply and remove a force;
- apply oscillating or reciprocating motion;
- expose to light;
- expose to moisture;
- stir;
- pour.

How will you do it?

What will cause your test to take place? Will it be an operator turning a handle or adding weights? Will it be an electric motor causing a turning force or movement? Will it be a pneumatic cylinder causing a pushing force? You have to decide.

What will you measure?

You may need to measure inputs, such as:

- loads, e.g. force, torque (turning force);
- number of operations, e.g. strokes or cycles;
- how long input is applied; physical conditions, e.g. light level, temperature, humidity;
- starting dimensions.

You may need to measure the effects of the test, such as when failure occurs. Is this at a particular load, after a certain number of strokes or cycles or under certain conditions? You might measure how long it takes for something to happen, the maximum loading achieved or change in the dimensions of the sample.

You may need to *describe* the effects of your test. This will involve careful observation of the sample or object at the beginning and end of the test. You might need to describe some of the following:

- scuffing
- piling
- creasing
- scratching
- necking
- changes in colour, flexibility or viscosity.

Will using a computer help?

If you design your testing so that the inputs and outputs can be measured as electrical signals, either digital or analogue, then using a computer can help.

Switches can provide digital information. Variable resistors, strain gauges, light-dependent resistors and thermistors can provide analogue information. You will need to use an interface to take the signals into the computer and a program to handle this information and display it in a way that makes sense – usually as a graph.

Body adornment

Thinking about what's possible

Here you can see the different types of body adornment that you might design and make. For each type of body adornment you should ask yourself the following questions.

- What sorts of design already exist?
- What ready-made parts can I use?
- What materials might I use?

- What will I need to measure?
- What are the important safety features?
- How will it affect clothing?

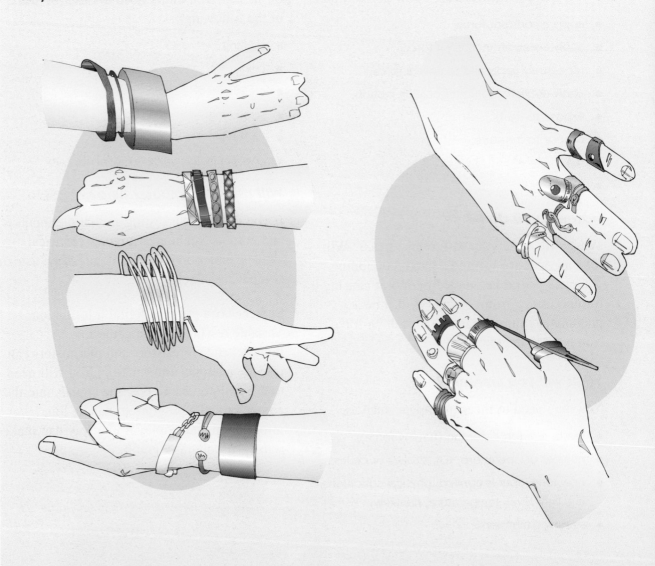

For your arms and hands …

LIRT1
LIRT2

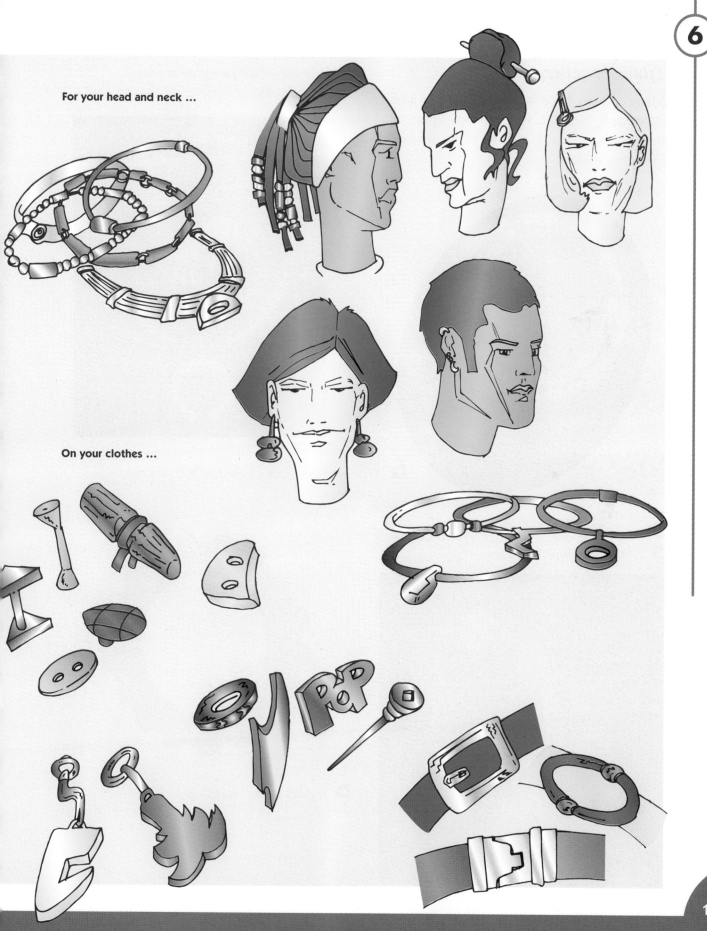

For your head and neck ...

On your clothes ...

Ethnic jewellery

Many cultures have strong traditions of body adornment. Here are some examples that you may be able to use as starting points for your own design ideas.

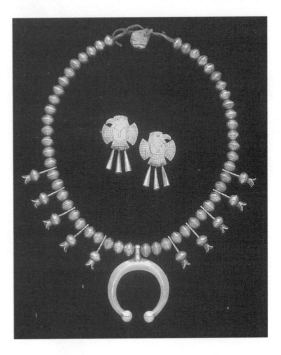

A Navajo woman's earrings made from silver (southern USA)

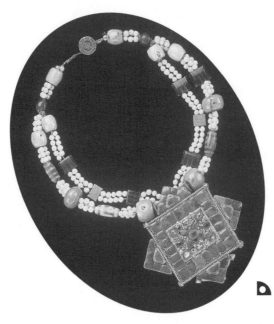

A noblewoman's necklace made from gold, turquoise, coral and pearls (Tibet)

A silver cloak pin (Chile)

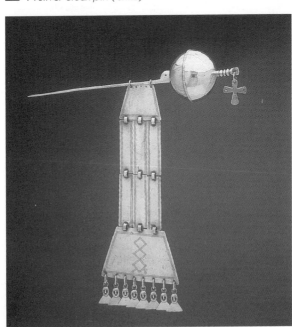

A chief's bronze neck ring (Zaire, Africa)

Influences on your designs

If your designs are to be successful you will need to take a wide range of influences into account. You can use the following questions to help you identify the factors that will influence your designs.

- **Social influences**

  What style is likely to be appropriate for the wearer – plain or fancy, cheap or expensive, abstract or representational, bright or dull, large or small?

  Is it a reward, or an award?

  Should it show that the wearer is wealthy or powerful?

  Should it show that the wearer holds a particular office?

- **Allegiance influences**

  Should it show that wearers belong to a particular group such as a fan club, a supporters' club or a sports' team?

  Should it indicate that wearers support a particular cause?

  Should it indicate religious belief?

- **Personal influences**

  Is it a token of affection?

  Is it used to show a relationship?

- **Functional influences**

  Will it be used to keep hair or clothes in place?

  Will it be used as a 'lucky charm'?

  Will it be used to relieve stress or worry?

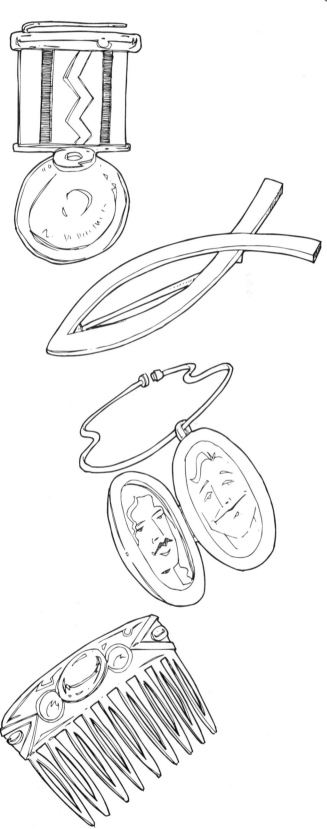

Findings

There are many components for body adornments that you should buy rather than make. These are usually called **findings**. Some are shown in the panel. They are usually available in a range of materials and cost more if they are made from solid gold or silver rather than plated. Your local jeweller will have these and other findings in stock.

Making your body adornment

You can use the flow chart below to help you produce a quality product.

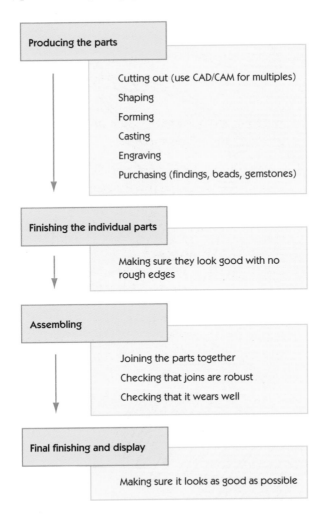

Producing the parts

Cutting out (use CAD/CAM for multiples)
Shaping
Forming
Casting
Engraving
Purchasing (findings, beads, gemstones)

Finishing the individual parts

Making sure they look good with no rough edges

Assembling

Joining the parts together
Checking that joins are robust
Checking that it wears well

Final finishing and display

Making sure it looks as good as possible

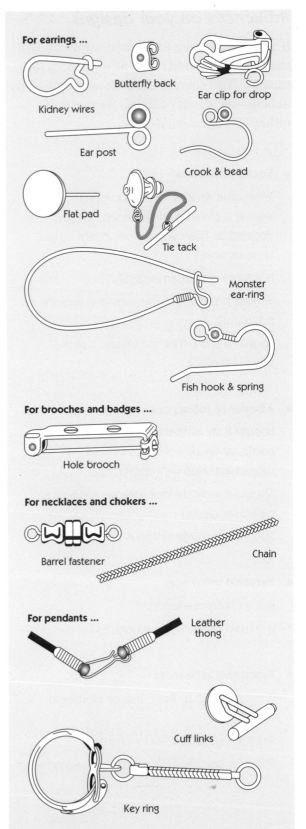

For earrings ...

Kidney wires

Butterfly back

Ear clip for drop

Ear post

Crook & bead

Flat pad

Tie tack

Monster ear-ring

Fish hook & spring

For brooches and badges ...

Hole brooch

For necklaces and chokers ...

Barrel fastener

Chain

For pendants ...

Leather thong

Cuff links

Key ring

Findings for body adornment

Designing to improve quality

It is unlikely that your design will be so good that it cannot be improved. Here are five areas that you should consider:

- manufacture;
- durability;
- maintenance and repair;
- disposal
- sustainability.

Designing for manufacture

Answering these questions will help you design a product that is easier to manufacture.

- **Can the number of parts be reduced?**
  Some parts may be unnecessary; other parts may be combined into a single part. The production of fewer parts will save on manufacturing time.

- **Are all the parts the simplest possible shape?**
  Simple shapes are easier and quicker to produce than complicated shapes.

- **Can some parts be bought ready made?**
  This saves on manufacturing time and equipment.

- **Can identical parts be produced more efficiently?**
  You should always consider using CAD/CAM. If this is inappropriate, develop ways of making several parts simultaneously.

- **Is assembly rapid and foolproof?**
  The fewer the parts and the fewer the fixings the more rapid the assembly. Redesigning parts so that they can fit together in one way only decreases assembly time and prevents mistakes.

- **Can any processes be eliminated or reduced in time?**
  Some parts will need to be finished to fine tolerances; for others this will not matter. Ensure that you are always working to an appropriate degree of accuracy. Remember, the greater the accuracy the longer the time taken.

Designing for durability

What parts are likely to wear out, break or become damaged during typical use?

How can the design of these parts be changed to make them last longer? Here are some possibilities.

- **Changing the shape of the part**
  Radius corners and edges are less likely to crack or snag.

- **Changing the thickness of the part**
  Thicker parts will be stronger.

- **Changing the material the part is made from**
  Some materials are stronger than others and will break less easily. Some materials are more resistant to corrosion and rotting than others.

- **Changing the finishes used**
  Give each part a protective finish suitable for the working conditions.

- **Decreasing friction between moving parts**
  Ensure that moving parts are well finished and that lubrication is provided where necessary. Include bearings that will reduce wear. Choose materials that run smoothly together.

Design guides – quality

Designing for maintenance and repair

Some parts are likely to wear out, break or become damaged during typical use.

The design of the product should be changed to achieve the following.

- **Ease of access**
  Access panels are one way of doing this. It is important that they are the right size, in the correct position and easy to remove and replace.

- **Ease of removal and replacement**
  The way that parts are held in place is important here. Parts force-fitted together are not helpful because they are difficult to prise apart. Keyways, grub screws and spring clips offer useful alternatives.

Designing for disposal

At the end of its useful life a product will usually be thrown away. It will become increasingly important for the design of the product to achieve the following.

- **Ease of dismantling**
  This will enable useful parts that have not worn out to be reclaimed and re-used.

- **Ease of material identification**
  This will enable materials to be recycled.

- **Ease of recycling**
  Some of the materials chosen for the design should be suitable for recycling.

Designing for sustainability

Many designers are now beginning to take a much broader view of the impact of their work on the world and its resources. They are looking at their designs with a view to minimising impact on the environment. This has some interesting consequences. For example one washing machine manufacturer is designing machines that can be upgraded as new improvements are developed. This means that customers will not have to buy a completely new washing machine when they become dissatisfied with the performance of their current model. So the large amounts of material that are used in the frame, washing drum and shell will have a much longer useful life. An improved machine will no longer need to be made from completetly new materials and parts.

This approach will be particularly applicable to complex products that use information technology in the way they work. The idea of upgrade is already common in computer software. It is likely to be applied to many of the following everyday products; motor cars, domestic heating systems, domestic machines such as washing machines and cookers, computer systems, music centres, and communication devices – telephones and faxes. For many designers this will mean rethinking their approach to design and adopting a modular approach.

Recall

Mechanisms can:

- change direction of movement, e.g. from clockwise to anticlockwise or from horizontal to vertical;

- change type of movement, e.g. from rotating to linear or from reciprocating to oscillating;

- alter axis of movement, e.g. from horizontal to vertical;

- increase speed and distance while output force is reduced;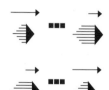

- increase output force while speed or distance is reduced;

- apply and maintain a force;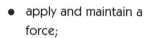

- transmit force and movement.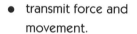

Here are some examples.

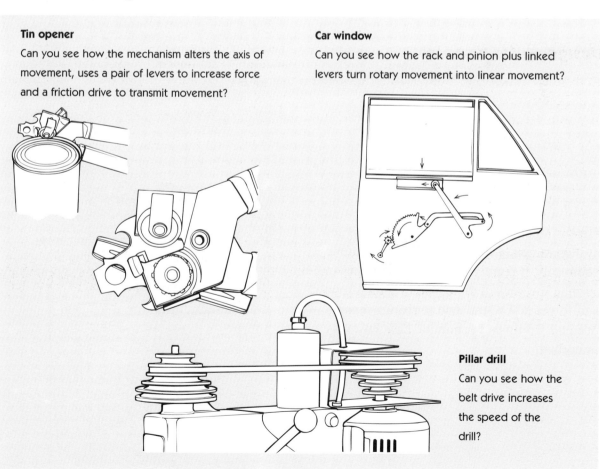

Tin opener

Can you see how the mechanism alters the axis of movement, uses a pair of levers to increase force and a friction drive to transmit movement?

Car window

Can you see how the rack and pinion plus linked levers turn rotary movement into linear movement?

Pillar drill

Can you see how the belt drive increases the speed of the drill?

Mechanical systems

Starting and stopping

Driver Chooser Chart

You can use this information to help you to choose the best driver for your mechanical system.

Type of driver	d.c. motor	clockwork motor	windmill/ wind turbine	waterwheel/ water turbine	pneumatic cylinder	solenoid	ramrod
Type of movement	↻	↻	↻	↻	or ⇄	or ⇄	or ⇄
Power source	battery	human muscle, energy stored in spring	wind	moving water	compressed air	battery	battery
Output force or torque	low	low–medium	low	medium	high	low	medium
Output speed	high	low	low–medium	low–medium	low–high	medium–high	low
Cost	£	££	£££*	££££*	££££	£££	£££
Other features	Can be reversed by reversing polarity	Runs for limited time in-between rewinds	Requires wind. Very low power for size	Can be reversed through gear box or pulley system	Fixed stroke. Only stops at one end or other. Reversed by spring (sac) or air (dac)	Fixed stroke. Only stops at one end or other. Reversed by switching off	Limited stroke. Can be stopped mid-stroke. Reversed by reversing polarity

* if home-made

Here is an example. Sarah was designing a system to gently rock a tray of photographic chemicals. It could be powered by battery or spring. She wanted to be able to switch it on and forget it. The tray had to rise 25 mm at one end, 15 times per minute. Low cost was important.

Sarah looked at the pros and cons of several drivers:

Driver	Pros	Cons
Pneumatic cylinder	Provides reciprocating motion ⟷	High cost needs comp.air
Solenoid	Provides reciprocating motion ⟷	high cost needs on/off circuit high current drain on battery
Ramrod	Provides reciprocating motion ⟷	high cost needs reversing switch high current drain on battery
DC motor	Low cost readily available ↻	Provides rotary motion, low force high speed high current drain on battery

To keep the cost low, she chose the d.c. motor, even though it produced rotary motion, not reciprocating motion.

She remembered two useful mechanisms:

- worm and wheel, to reduce speed and increase force;
- cam and follower, to change rotary motion to reciprocating motion.

She checked the Mechanisms Chooser Chart, page 158–159, to make sure.

Sarah's initial choice was a d.c. motor, driving a worm and wheel, driving a cam and follower. This was low cost, and would raise and drop the tray without the need for splash-proof reversing switches.

Testing the initial choice

If you know exactly what force/torque, speed, etc. you need, you may be able to choose the right driver from a catalogue. Often it will be cheaper to use a supplier who doesn't specify all the technical details. In this case, it is useful to set up a test rig.

Sarah set up this rig to test the motors that her teacher had. By trial and error, she found a motor and worm wheel which gave enough force and the right speed. She knew that this choice of driver would work.

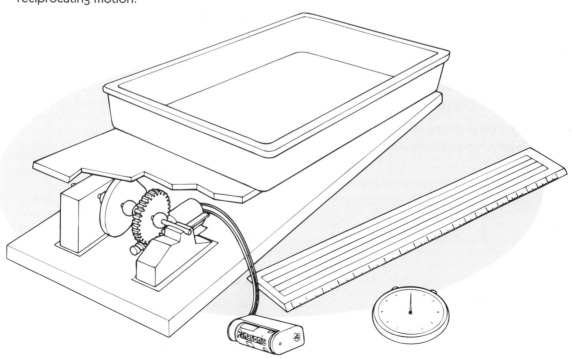

Transmission

Shafts

A shaft transmits rotation along its own axis, as in this sewing machine.

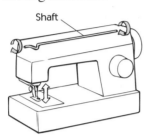

Shaft

Universal joints are used to transmit rotation between shafts that meet at an angle as in this socket set.

Design factors

- A shaft must be stiff enough to resist twisting and sagging.
- The angle between shafts must be less than 20°.
- If a single universal joint is used the output shaft will not rotate at a constant speed even though the input shaft is rotating at a constant speed.
- For a constant speed input and one universal joint, the output speed is not constant. To achieve a constant output speed you need to use two universal joints as shown.
- For the transmission of low forces you can use rubber tubing or springs for universal joints.

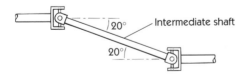

20°   Intermediate shaft

20°

Input   Output
15°   Intermediate   15°

Belts and chains

Belts and chains transmit rotation from one axle or shaft to a parallel axle or shaft, as shown in these examples.

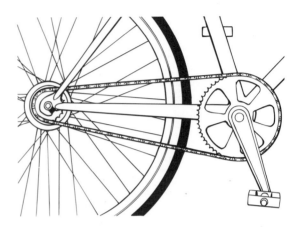

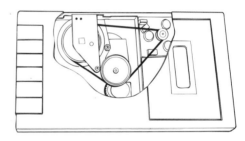

Design factors

Chains and toothed belts:

- don't slip;
- give positive drive;
- transmit larger forces than belts;
- require more accurate spacing and alignment than belts;
- are more expensive and noisier than belts;
- often need lubrication.

Plain belts:

- slip under excessive loads;
- rarely need lubrication.

Rods and links

Rods and links transmit forces from cranks and levers, as shown in the aquarium pump and the squeezy mop.

Design factors

- Rods and links must be stiff enough to resist buckling when in compression.

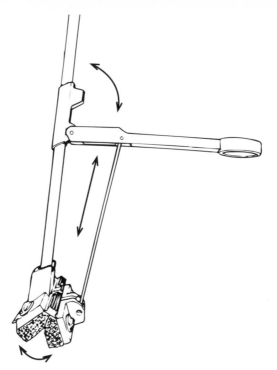

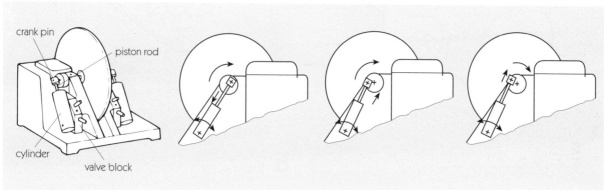

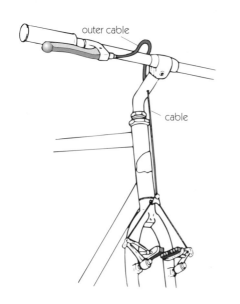

Cables

Cables transmit pulling forces along straight or curved paths, as shown in this bicycle brake.

Design factors

- Cables cannot provide pushing forces and any reversal of the input movement requires an additional force, usually provided by a spring or gravity.

Other functions

Many transmission systems will include belt or gear systems which will enable them to change direction of movement, alter axis of movement and change speed.

Clutches

Clutches connect or disconnect two shafts that are in line with one another. Positive clutches are used for shafts at rest; slip clutches are used when one or both of the shafts are rotating. For shafts that are parallel you can use a belt tensioner as a clutch.

Design factors

- Usually one shaft must be able to slide as well as rotate so that it can engage and disengage with the other shaft.
- Sometimes neither shaft slides but part of the clutch can slide on one of the shafts.
- You will need to design a means of moving the sliding parts and you may need to use a spring to hold the clutch parts together.

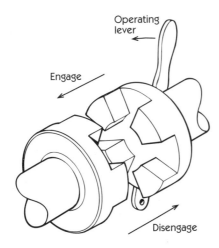

▶ A dog clutch

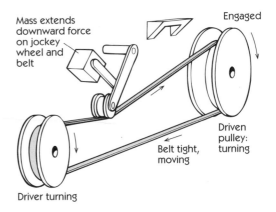

▶ Belt tensioner

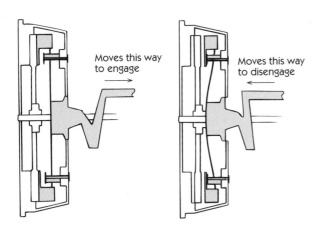

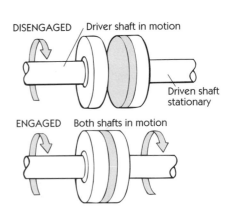

▶ Friction plate clutch

Brakes and governors

Brakes

Brakes bring moving parts to rest, keep moving parts stationary (or parked) and reduce the speed of moving parts. They all work by pushing a stationary surface against the surface of a moving part. Friction provides the slowing force and heat is generated.

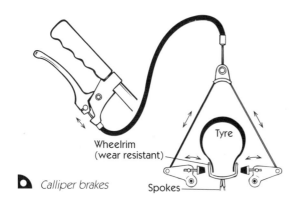

▶ *Calliper brakes*

Wheelrim (wear resistant)
Tyre
Spokes

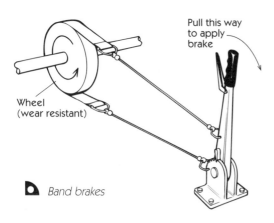

Pull this way to apply brake

Wheel (wear resistant)

▶ *Band brakes*

Design factors

- Choose materials which don't soften, melt or distort when they get hot.
- You will need to choose how force will be applied to put the brake on *and* to take it off.
- You will need to decide if your brake needs to stay on when the operator moves away, like a car handbrake. If so you will need a mechanism that applies and maintains force (see page 155).

Governors

Governors smooth out variations in speed and prevent parts from moving too fast.

A **centrifugal governor** is used to control the speed of a rotating shaft. The masses move outwards as the shaft rotates more quickly. As they move outwards it takes more energy and force to move them so the rotating shaft slows down. As the shaft slows down the masses move inwards and require less energy and force to move them, so the rotating shaft speeds up.

For any one power input a given pair of masses will maintain a particular speed. The governor can control the power used to move the shaft by connecting the governor collar to a control device – a fuel valve, steam valve or electrical resistor. As the collar rises, the amount of fuel, steam or electrical current is reduced and vice versa.

Design factors

- You will need to experiment with both the geometry and the masses of a centrifugal governor in order to get the speed control that you need.

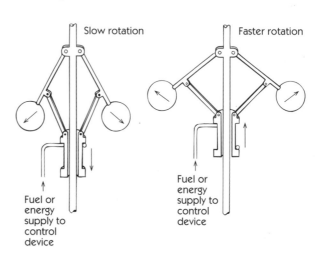

Slow rotation   Faster rotation

Fuel or energy supply to control device

Fuel or energy supply to control device

▶ *Centrifugal governor*

Changing force and movement

Levers

Levers can be used to:

- increase force and decrease speed or distance travelled, as in a crowbar or wheelbarrow;

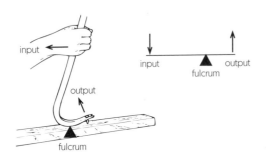

- increase speed or distance travelled and decrease force, as in a pair of tweezers;

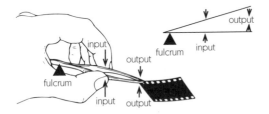

- change direction, as in a bell crank.

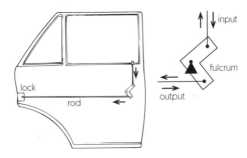

Design factors

- To increase output force, distance of effort from fulcrum must be greater than distance of load from fulcrum.

- To increase speed or distance travelled, distance of effort from fulcrum must be less than distance of load from fulcrum.

- To change direction, fulcrum must be between load and effort.

Linkage systems

3-bar linkages are rigid, but by making one of the bars adjustable in length in some way this linkage system becomes a useful mechanism. Here are some examples.

- The framework of a deckchair is an adjustable 3-bar linkage system. It forms a load-bearing structure that can be adjusted.

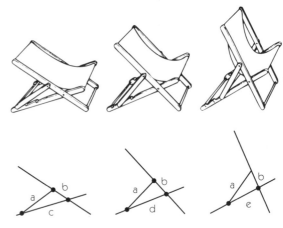

- The car jack contains an adjustable 3-bar linkage system. It forms a mechanism for raising and lowering heavy loads.

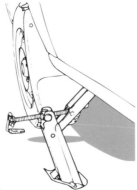

Design factors

- You will need to design a means of keeping the adjustable bar in a fixed position for a rigid structure.

4-bar linkages are not rigid and have many applications, for example:

- keeping moving parts parallel, as in the lamp below;
- providing fold-flat structures, as in this double-glazed window hinge;
- converting rotary movement to reciprocating or oscillating movement, as in this mechanical hacksaw.

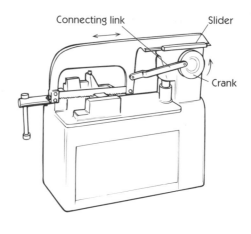

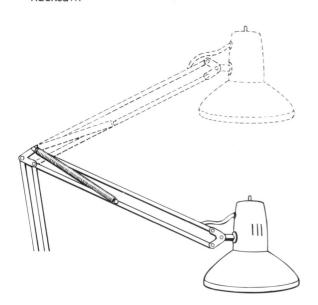

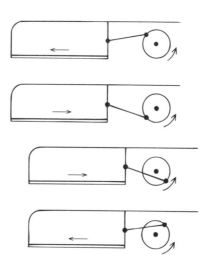

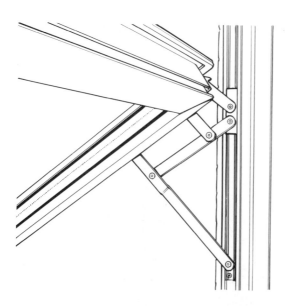

Design factors

- When used to keep moving parts parallel, you will need to design the linkages to form a parallelogram; i.e. opposite sides equal in length but each pair different in length.

- In fold-flat structures, the linkages need not form a parallelogram but the total length of one pair of sides must equal the total length of the other pair.

- To enable a complete rotation of the crank arm when converting rotary to reciprocating or oscillating movement, you will need to ensure that the combined length of the shortest and longest bars is less than or equal to the combined length of the other two bars.

Cams

A cam can be used to convert rotating movement to either oscillating or reciprocating movement.

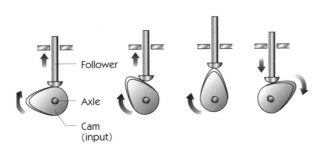

- Follower
- Axle
- Cam (input)

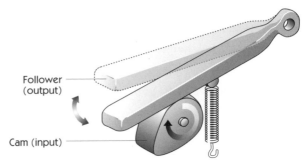

- Follower (output)
- Cam (input)

 Cam producing reciprocating movement

 Cam producing oscillating movement

Design factors

- By changing the profile (shape) of the cam you can control the way the follower moves, as shown below.

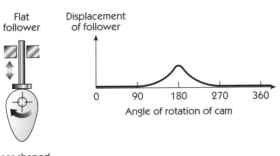

Flat follower

Displacement of follower

0   90   180   270   360
Angle of rotation of cam

Pear-shaped cam (intermittent)

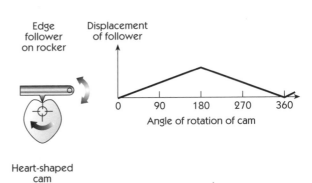

Edge follower on rocker

Displacement of follower

0   90   180   270   360
Angle of rotation of cam

Heart-shaped cam (uniform velocity)

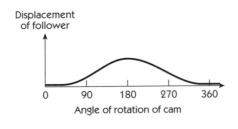

Roller follower

Displacement of follower

0   90   180   270   360
Angle of rotation of cam

Eccentric cam

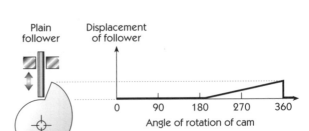

Plain follower

Displacement of follower

0   90   180   270   360
Angle of rotation of cam

Snail cam (drop cam)

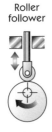

The Geneva mechanism combines a cam and follower with a peg and slot as shown. For a constant speed rotation input, it produces intermittent part rotation output. It has many applications, for example in moving toys and automata.

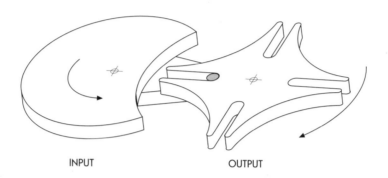

INPUT   OUTPUT

Design factors

● You will need to position the shafts for the cam and the follower precisely and ensure that the peg fits smoothly into the slots.

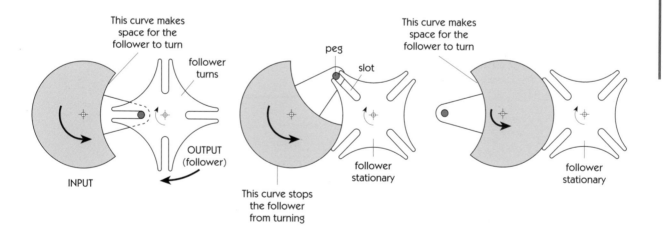

This curve makes space for the follower to turn

follower turns

OUTPUT (follower)

INPUT

This curve stops the follower from turning

peg

slot

follower stationary

This curve makes space for the follower to turn

follower stationary

Gears

Gears can be used for a variety of purposes, for example:

- change the type of motion;

- change the direction of rotation;

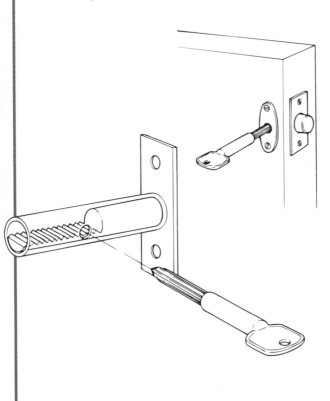

FORWARD

spindle gear

forward

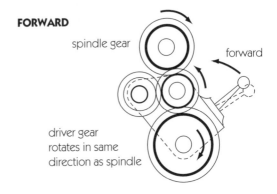

driver gear
rotates in same
direction as spindle

REVERSE

spindle gear

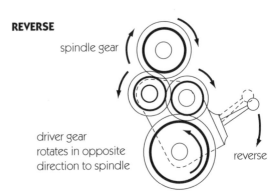

driver gear
rotates in opposite
direction to spindle

reverse

- alter the axis of rotation;

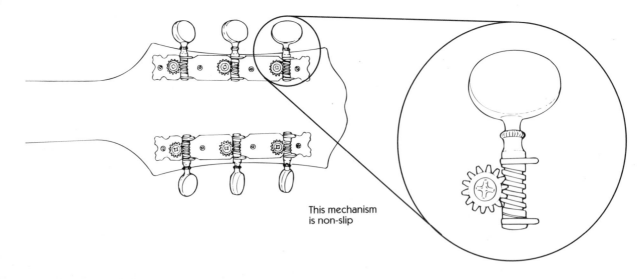

This mechanism
is non-slip

- increase output force (and decrease output speed);

- increase output speed of rotation (and decrease output force).

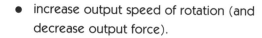

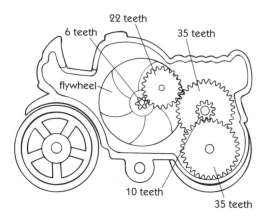

"To store enough energy to make this toy go, the flywheel must turn very fast. When the user "scoots" the toy along the ground, the gears increase the speed, to make the flywheel go over 20 times faster than the back wheel"

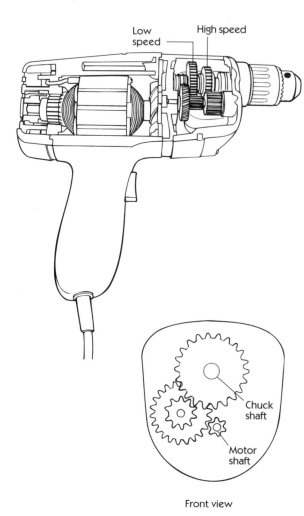

Front view

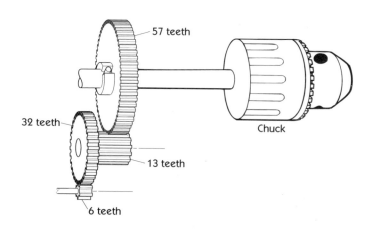

Design factors

- To increase speed and reduce force (gearing up) the output gear must have more teeth than the input gear.

- To increase force and reduce speed (gearing down) the output gear must have less teeth than the input gear.

- It is important to have the correct distance between the shafts. If too close together and the gears are hard to turn, wear more quickly and may jam; if too far apart the gears may slip.

- Metal gears may need lubrication to reduce friction, wear and noise.

Holding, supporting, attaching and adjusting

For supporting moving parts

Use bearings to provide support, reduce friction, reduce wear and confine wear to cheap materials.

For small to medium forces

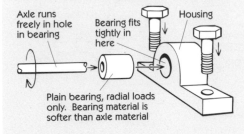

Axle runs freely in hole in bearing

Bearing fits tightly in here

Housing

Plain bearing, radial loads only. Bearing material is softer than axle material

Some plain bearings are made in two halves ("shells") to aid assembly

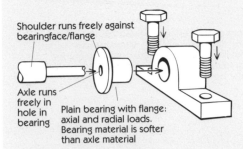

Shoulder runs freely against bearingface/flange

Axle runs freely in hole in bearing

Plain bearing with flange: axial and radial loads. Bearing material is softer than axle material

For medium to high forces

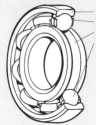

Outer race
Cage
Inner race
Ball bearing

Single row ball bearing – Radial forces Small axial forces

Taper roller bearing – High radial forces High axial forces Used in pairs

For attaching moving parts to stationary parts

Use pin joints. Take care to prevent overtightening or working loose.

For small forces

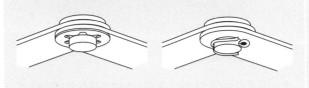

For larger forces

For attaching drivers

For small forces

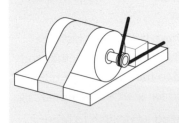

For low to medium forces

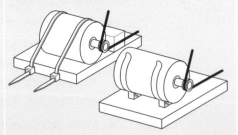

For larger forces

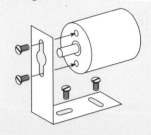

Holding, supporting, attaching and adjusting

For attaching to transmit rotating force (torque)

Shafts end-to-end

low torque

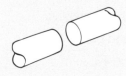

low to medium torque

medium to high torque

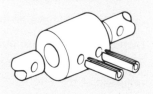

Components on shafts

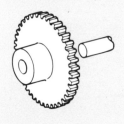

low torque

low to medium torque

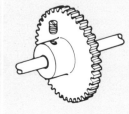

medium to high torque

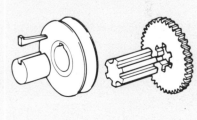

For adjusting

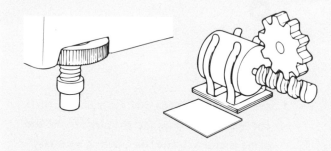
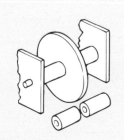
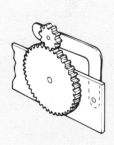

For holding

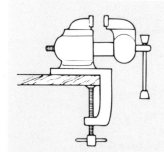
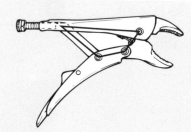
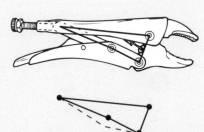

Assembly guide

These guidelines will help you assemble mechanical systems.

Before starting:

- identify all the parts;
- check that you have all the parts;
- plan the order of assembly so that you can easily access parts that have to be added or adjusted at the end.

When assembling:

- beware of over tightening (breakages);
- align and adjust before final tightening;
- check for missing or unused parts;
- test each subsystem as you go.

Here is an example showing how Ruth assembled this conveyor belt.

1 She checked all the parts against her component chart.

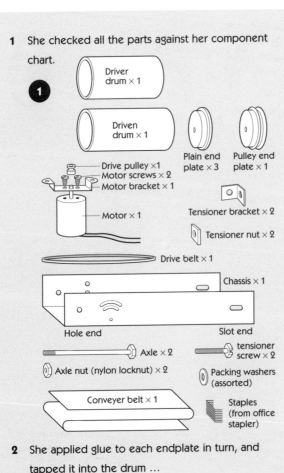

1
Driver drum × 1
Driven drum × 1
Plain end plate × 3
Pulley end plate × 1
Drive pulley × 1
Motor screws × 2
Motor bracket × 1
Motor × 1
Tensioner bracket × 2
Tensioner nut × 2
Drive belt × 1
Chassis × 1
Hole end
Slot end
Axle × 2
tensioner screw × 2
Axle nut (nylon locknut) × 2
Packing washers (assorted)
Conveyer belt × 1
Staples (from office stapler)

2 She applied glue to each endplate in turn, and tapped it into the drum …

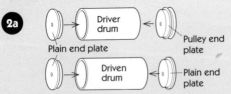

2a
Plain end plate → Driver drum ← Pulley end plate
Driven drum ← Plain end plate

She glued the square nuts onto the tensioner brackets …

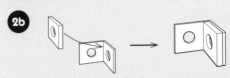

2b

and screwed the motor to its bracket, then tapped the pulley onto the motor shaft.

2c

3 She hung the drive belt over the drum. She fitted one axle and the drum with the pulley to the end of the chassis with axle holes, putting a standard washer between the chassis and the pulley endplate.

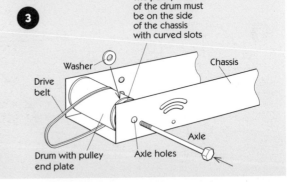

3

The pulley end of the drum must be on the side of the chassis with curved slots

Washer
Chassis
Drive belt
Axle
Drum with pulley end plate
Axle holes

4   She used a feeler gauge to measure the space between the endplate and the chassis. This is called **end float**. She added special thin washers to the gap until it was only 0.5 mm. Then she screwed a nylon locknut onto the end of the axle. She tightened the nut until it started to 'pinch' the drum, then she loosened it slightly until the drum turned freely.

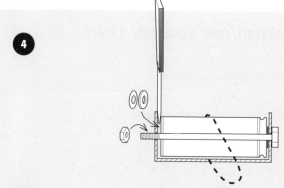

5   She fitted the motor through the hole in the chassis, and used a long screwdriver to tighten the motor bracket screws.

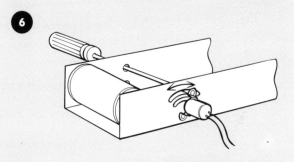

Screwdriver holes

6   She connected the motor to a battery, and adjusted the motor bracket until the belt was tight and the drum turned smoothly.

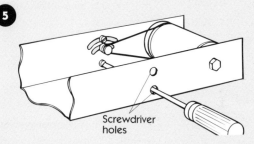

7   Using a large elastic band to hold the tensioner brackets in place, she screwed the tensioners into the brackets …

8   Then fitted the other drum and axle, adjusted the end float and tightened the other locknut (as in 4). She slackened the tensioners right off.

9   She looped the conveyor belt material around both drums, and stapled the ends together.

10 Finally, she tested the conveyor belt using the sorts of objects it was intended to convey. She adjusted the tensioner so that the belt didn't slip.

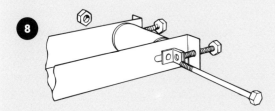

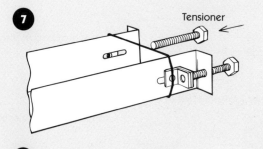

Tensioner

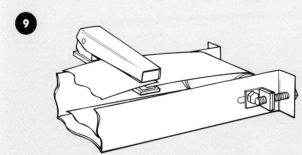

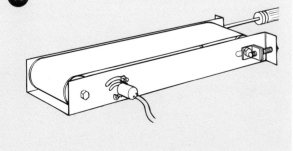

Mechanical systems

Mechanisms Chooser Chart

For this mechanical function:		You can use:
Changing the type of movement		
	from linear to rotating	wheel and axle, rack and pinion, screw thread, rope and pulley, chain and sprocket
	from rotating to linear	wheel and axle, rack and pinion, screw thread, rope and pulley, chain and sprocket
	from rotating to reciprocating	crank, link and slider (4-bar linkage), cam and slide follower
	from rotating to oscillating	crank, link and lever (4-bar linkage), cam and lever follower, peg and slot
	from rotating to intermittent rotating	Geneva wheel (a cam plus peg and slot)
	from reciprocating to rotating	crank, link and slider (4-bar linkage)
	from reciprocating to oscillating	wheel and axle, rack and pinion, crank and slider (4-bar linkage)
	from oscillating to rotating	crank link and lever (4-bar linkage), peg and slot
Changing the direction of movement		
	from clockwise to anticlockwise	gears, belt and pulley
	from left to right	levers, linked levers, rope and pulley
	from horizontal to vertical	levers, linked levers, rope and pulley

Mechanisms Chooser Chart

For this mechanical function:		You can use:
Changing the axis of rotation		
		bevel gears, flexible couplings, worm and wheel, belt and pulley
Increasing output force and decreasing speed		
	With parts rotating or oscillating:	gears, bevel gears, worm and wheel, wheel and axle, belt and pulley, chain and sprocket
	With parts reciprocating or moving in a straight line:	rope and pulley, levers, linked levers
Increasing output speed and decreasing force		
	With parts rotating or oscillating:	gears, bevel gears, wheel and axle, belt and pulley, chain and sprocket
	With parts reciprocating or moving in a straight line:	levers, linked levers
Applying and maintaining a force		
		screw thread, worm and wheel, cam, rope and pulley, toggle (4-bar linkage), spring, brake system
Transmitting force and movement		
	linear or reciprocating movement	linked levers, rods
	rotating or oscillating movement	belt and pulley, chain and sprocket, linked levers, shafts and couplings

Pneumatics

Pneumatic circuits work by means of compressed air. Here are two examples to introduce the components used to make a pneumatic circuit.

Putting tops on milk bottles

Here the pneumatic circuit is being used to produce linear movement and force. The piston moves up and down forcing the foil over the top of the bottle.

Opening and closing bus doors

The pneumatic circuit is being used to produce movement and force through the arc of a circle. The piston still moves in and out in a straight line but is connected to a lever system which causes the door to swing open and closed.

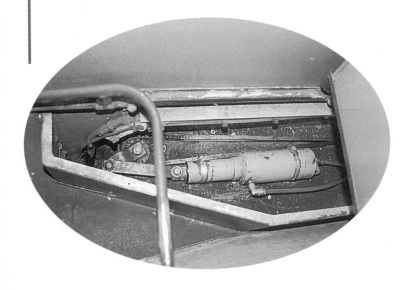

From these examples you should see that a pneumatic circuit requires the following components:

- compressor;
- receiver;
- valves;
- cylinders with pistons;
- air lines;
- actuators.

Making things happen with pneumatics

Operating a stamp

This simple circuit stamps when the push button is pressed. When the button is released the return spring pushes the piston back to its original position. The circuit diagram using symbols is also shown.

Pneumatic components used in operating a stamp			
For input (actuators)	For control	For sensing	For output
push button	3-port valve		single-acting cylinder

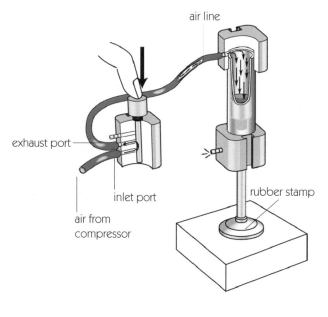

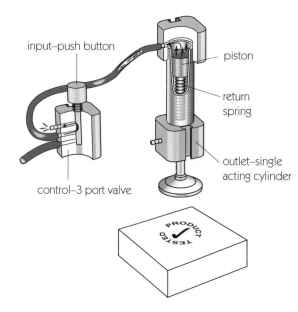

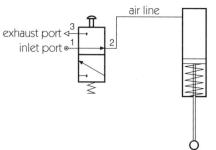

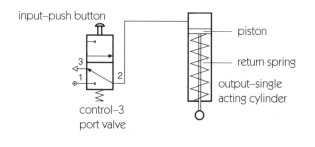

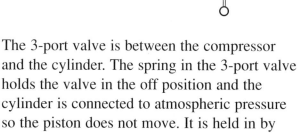

The 3-port valve is between the compressor and the cylinder. The spring in the 3-port valve holds the valve in the off position and the cylinder is connected to atmospheric pressure so the piston does not move. It is held in by the return spring.

When the button is pressed the valve connects the cylinder to the compressed air supply. This overcomes the return spring, the piston is pushed into the out position and the carton is stamped.

Making it safe

There is a serious hazard in this simple circuit – it would be very easy for the operators to 'stamp' one of their own hands. The arrangement can be made safe by putting in another 3-port valve in series with the first. This creates an AND gate. The air must go through first one valve AND then the other for the stamp to work. The operator has to use both hands to operate the valves so the risk is controlled.

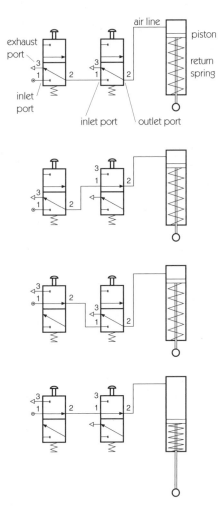

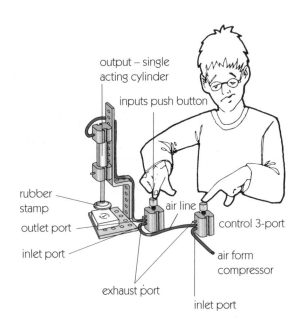

Pneumatic components used in operating a stamp			
For input (actuators)	For control	For sensing	For output
push button roller trip	3-port valve		single-acting cylinder

Ensuring the position of the box

The circuit diagram below includes a third 3-port valve to ensure that the stamp only works when the box is in the correct position for stamping.

The stamp only works when all three buttons are pushed at the same time – two for safe operating and one to ensure the box is in the right place.

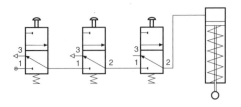

Simple textile tester

In this pneumatic circuit the designer chose to use a double-acting cylinder and a 5-port valve with a toggle switch.

When the toggle is as shown, the 5-port valve directs compressed air into the cylinder to make the piston move forwards.

When the toggle is switched, the 5-port valve directs compressed air into the cylinder to make the piston move backwards.

Pneumatic components used in simple textile tester			
For input (actuators)	For control	For sensing	For output
toggle	5-port valve		double-acting cylinder

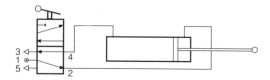

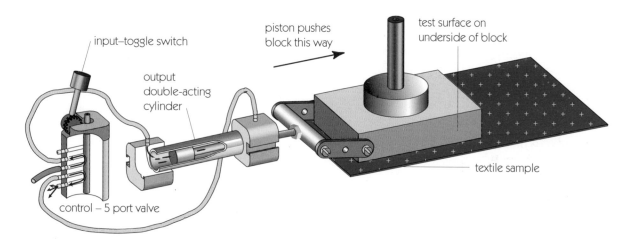

piston pushes block this way

test surface on underside of block

input–toggle switch

output double-acting cylinder

textile sample

control – 5 port valve

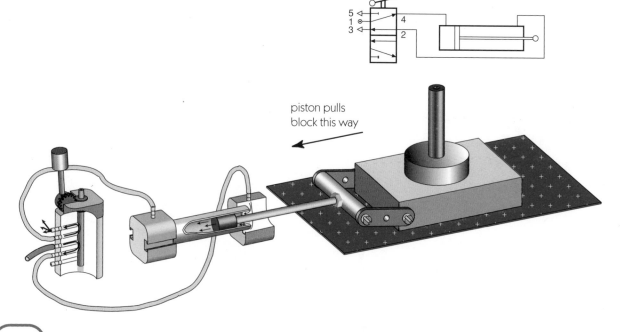

piston pulls block this way

An automatic textile tester

To produce an automatic textile tester which would go backwards and forwards by itself, the designer needed to find a way of sensing when the test block was at the ends of its stroke and then reversing the direction of the piston. She did this by using a roller trip 3-port valve and a push button 3-port valve as sensors and a 5-port valve activated by signal air.

At the end of the return stroke the roller trip on this 3-port valve is pressed. This connects signal air to the 5-port valve which switches the valves so that the piston moves forwards.

At the end of the forward stroke the block pushes the button on this 3-port valve which connects signal air to the other end of the 5-port valve. This switches the valve so that the piston moves backwards.

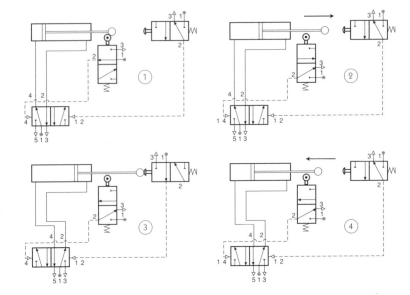

Pneumatic components used in an automatic textile tester			
For input (actuators)	For control	For sensing	For output
signal air	5-port valve	push button roller trip	double-acting cylinder

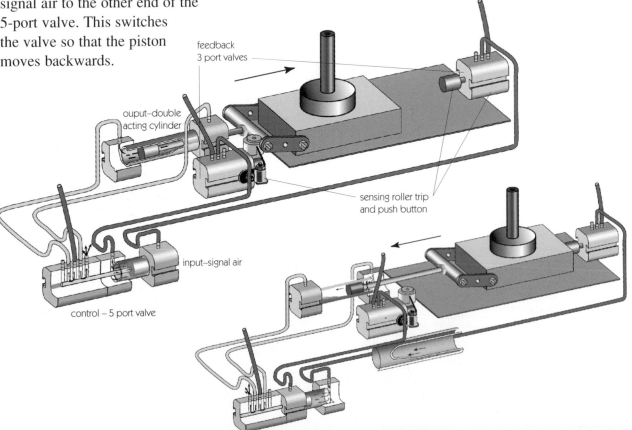

feedback
3 port valves

ouput–double
acting cylinder

sensing roller trip
and push button

input–signal air

control – 5 port valve

The snack bar wrapper

In this pneumatic circuit the piston is used to dispense snack bars onto their wrappers. The piston has to move out slowly to avoid forcing the snack bar off the slippery wrapping paper. It then has to pause and move in quickly. This quick return stroke saves time. The pneumatic circuit shown here meets these requirements.

Pneumatic components used in a snack bar wrapper			
For input (actuators)	For control	For sensing	For output
toggle	5-port valve	push button	double-acting cylinder
signal air	non-return valve	roller trip	

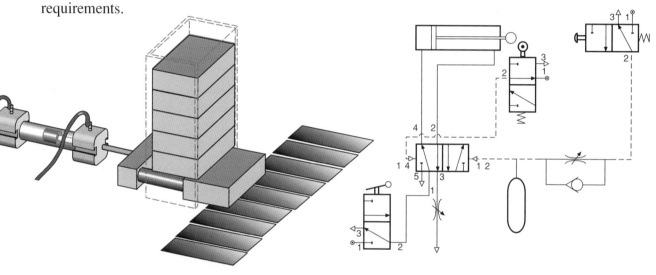

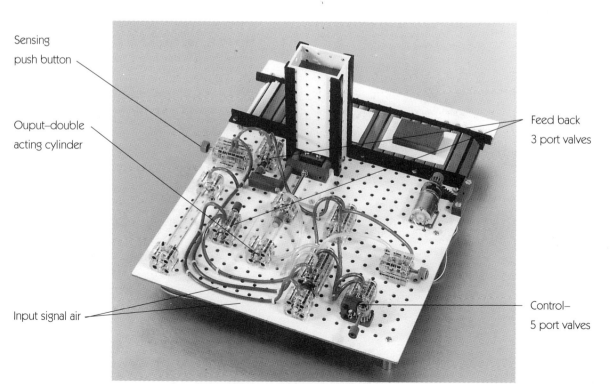

Sensing push button

Ouput–double acting cylinder

Input signal air

Feed back 3 port valves

Control– 5 port valves

Mechanical systems

Pneumatics Chooser Chart

Component	Function	Applications
For input – actuators		
toggle switch	to switch the valves in a 3-port or 5-port valve	hand operation of control valves
push button	to switch the valves in a 3-port or 5-port valve	hand or foot operation of control valves
plunger	to switch the valves in a 3-port or 5-port valve	mechanical operation (by pedal, lever, piston rod or cam) of control valves
roller trip	to switch the valves in a 3-port or 5-port valve	mechanical operation (by pedal, lever, piston rod or cam) of control valves
signal air	to switch the valves in a 3-port or 5-port valve	to provide for remote switching
solenoid	to switch the valves in a 3-port or 5-port valve	electrical operation of control valves
For control		
3-port valve	to control the flow of air to a single connection in a component by switching the connection from compressed air to exhaust or exhaust to compressed air	controlling a single-acting cylinder two valves in series make an AND gate two valves in parallel with a shuttle valve make an OR gate one valve with air supply connected to port 3 and exhaust connected to port 1 makes a NOT gate
5-port valve	to control the flow of air to two connections at the same time. One connection is switched from compressed air to exhaust as the other connection is switched from exhaust to compressed air	controlling a double-acting cylinder latching a double-acting cylinder

Component	Function	Applications
For control		
non-return valve	to allow air to pass in one direction only	used in parallel with a flow restrictor to give different speed forward and reverse strokes on a double-acting cylinder
shuttle valve	to allow air to take only one of the two routes through a T-junction	used with a 3-port valve to produce an OR gate
flow restrictor	to control the rate of air flow	on its own to control piston speed in both directions, with a non-return valve to give different speed forward and reverse strokes, with a reservoir to control length of time delay
reservoir	to produce time delays	used in series with a flow restrictor to give different lengths of time delay
For sensing		
push button	to switch the valves in a 3-port or 5-port valve	used to detect the position of objects or pistons
plunger	to switch the valves in a 3-port or 5-port valve	used to detect the position of objects or pistons
roller trip	to switch the valves in a 3-port or 5-port valve	used to detect the position of objects or pistons
reed switch		used with a solenoid activated 3-port or 5-port valve to detect the position of magnetized objects or pistons
micro-switch		Used with a solenoid activated 3-port or 5-port valve to detect the position of moving objects or pistons
For output		
single-acting cylinder	produces linear force and movement in **one** direction along a line for fixed distance (the stroke) reverse movement provided by return spring	to provide force and movement at the output of the system, e.g. controlled with a 3-port valve
double-acting cylinder	produces linear force and movement in *two* directions along a line for a fixed distance (the stroke)	to provide force and movement at the output of the system, e.g. controlled with a 5-port valve

Calculations for machines

Levers

You can calculate the effort required to move a load with a lever if you know the position of the fulcrum and the size of the lever. It is important to take the direction of the applied effort into account. If it is not perpendicular to the lever then the effect on the load is reduced.

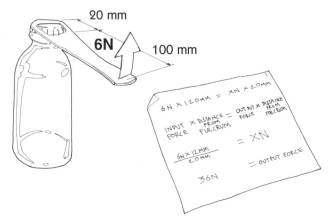

20 mm

6N

100 mm

$6N \times 120MM = XN \times 20MM$

INPUT × DISTANCE = OUTPUT × DISTANCE
FORCE   FROM   FORCE   FROM
        FULCRUM   FULCRUM

$\frac{6N \times 12MM}{20MM}$ = XN

= OUTPUT FORCE

36N

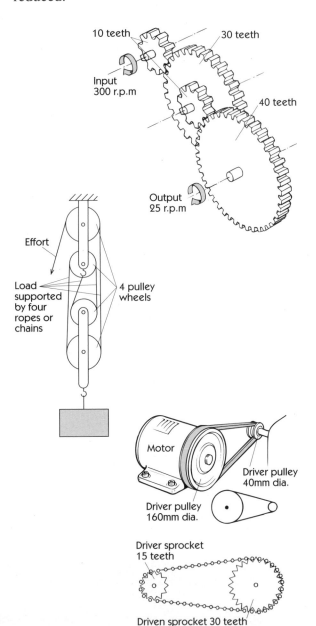

10 teeth

30 teeth

Input
300 r.p.m

40 teeth

Output
25 r.p.m

Effort

Load
supported
by four
ropes or
chains

4 pulley
wheels

Motor

Driver pulley
40mm dia.

Driver pulley
160mm dia.

Driver sprocket
15 teeth

Driven sprocket 30 teeth

Gear trains

You can calculate the output speed for a compound gear train if you know the input speed and the number of teeth on each gear.

You can estimate the change in output force like this.

> If the gear train doubles the output speed then the output force will be at most half that of the input force. If the gear train halves the output speed then the output force will be at most double the input force.

Pulleys

To calculate the effort required to move a load with a pulley system you need to know the number of pulleys in the system.

Belts and chains

You can calculate the output speed for a belt or chain drive if you know the input speed and the size of each pulley or chain wheel.

Efficiency

Helen knew she could apply a force of 200 N. She wanted to lift a small engine weighing 600 N. She set up the pulley system with a mechanical advantage of 3 but she found that she couldn't lift the engine. She checked her input force with a force meter – nothing wrong there, but she had to put an extra pulley wheel in the system to be able to lift the engine.

Helen had discovered an important principle about the efficiency of a machine. The efficiency of a machine is given as:

$$\frac{\text{velocity ratio}}{\text{mechanical advantage}}$$

$$\text{where } (VR) = \frac{\text{distance moved by input}}{\text{distance moved by output}}$$

$$\text{and } (MA) = \frac{\text{output load}}{\text{input load}}$$

(VR = velocity ratio and
MA = mechanical advantage)

Looking at her initial pulley system, $VR = 3$ and $MA = 3$ so it appeared that this pulley system should be 100 per cent efficient. However, Helen had ignored friction. The heavy load of the engine increased the friction between the rope and the pulley, and between the pulley block bearings and their axles. Her first calculation did not take these forces into account. By putting another pulley block into the system she increased the mechanical advantage. Helen could then *just* lift the engine block. Theoretically, she should have been able to lift much more, but the actual efficiency of the machine was reduced by friction.

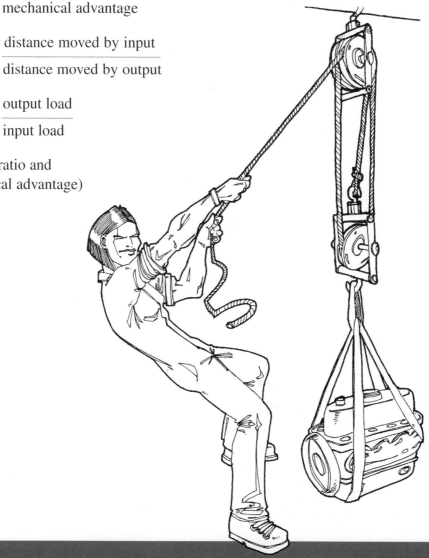

Power

You can calculate the power of a mechanical system by measuring the time it takes for a known weight (measured in newtons) to be lifted through one metre.

The work done in lifting the weight is given by:

force $\times$ distance moved

So in the case of the windmill

work done   = 10 N $\times$ 1 m
       = 10 Nm (newton metres)
       = 10 J (joules)

The power of the system is given by:

$$\frac{\text{work done}}{\text{time taken (s)}}$$

This is the amount of work done in one second.

In the case of the windmill:

power   = $\dfrac{10\ J}{1\ s}$

    = 10 J/s
    = 10 W (1 watt = 1 joule per second)

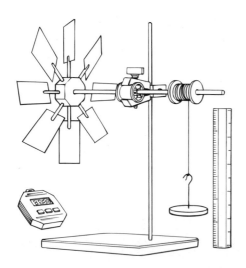

Power efficiency

The windmill is shown connected to a dynamo and lighting a bulb. The current passing through the bulb can be measured easily using an ammeter in series with the bulb, and the voltage across the bulb can be measured using a voltmeter in parallel with the bulb. So in the case of the windmill-driven dynamo, the power of the system is given by:

current (in amps) $\times$ voltage (in volts)

= 1 A $\times$ 1 V

= 1 coulomb per second $\times$ 1 joule per coulomb

= 1 J/s

= 1 W

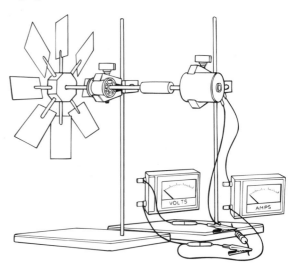

Efficiency

The efficiency of any system is the ratio of the power output to the power input. So the efficiency of the windmill-powered dynamo is given by:

1 W/10 W = 0.1

This is usually shown as a percentage:
0.1 $\times$ 100 = 10%

Tension and compression

The arch bridge and the aerial ropeway in the panels below were designed to resist specific loads and forces. The arch has to resist the load of the tractor. To support the tractor the stones in the arch are squeezed together. This squeeze force is called a **compressive force**, because to resist the load of the tractor the stones in the arch become compressed.

Only materials which are good at resisting compressive forces can be used for building this type of bridge. Since stone is strong in compression it is a good choice of material to use.

Now think about the aerial ropeway. The person's weight (the load) stretches the rope, pulling it tight. When the rope is tight, it is in **tension**. This is unlike the stones in the arch which are compressed.

Notice how the forces of the load and the resistance to the load are shown in these diagrams.

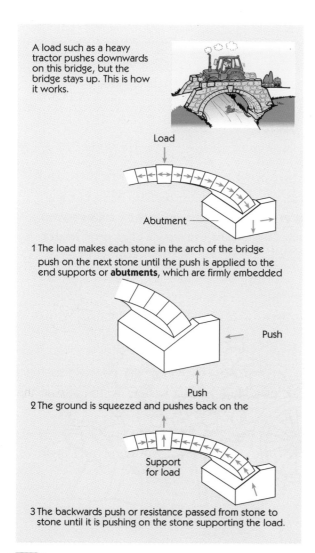

A load such as a heavy tractor pushes downwards on this bridge, but the bridge stays up. This is how it works.

Load

Abutment

1 The load makes each stone in the arch of the bridge push on the next stone until the push is applied to the end supports or **abutments**, which are firmly embedded

Push

Push

2 The ground is squeezed and pushes back on the

Support for load

3 The backwards push or resistance passed from stone to stone until it is pushing on the stone supporting the load.

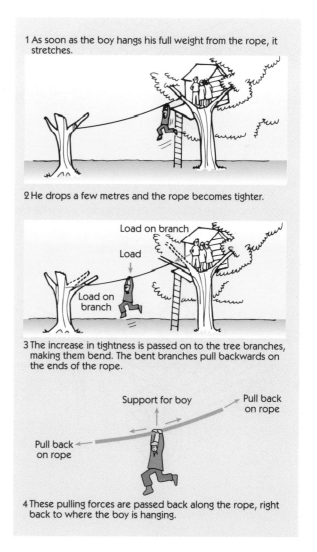

1 As soon as the boy hangs his full weight from the rope, it stretches.

2 He drops a few metres and the rope becomes tighter.

Load on branch

Load

Load on branch

3 The increase in tightness is passed on to the tree branches, making them bend. The bent branches pull backwards on the ends of the rope.

Support for boy

Pull back on rope

Pull back on rope

4 These pulling forces are passed back along the rope, right back to where the boy is hanging.

Stability and balance

Why has this apple cart toppled over? The cart was strong enough to resist the load but the positioning of the apples meant that there was nothing to counterbalance the load. It is like having all the apples on one side of a balance – if there is no weight on the other side of the scales they would tip up just like the apple cart.

The cart could be redesigned to make it stable no matter where the apples are loaded.

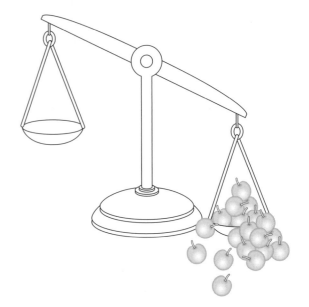

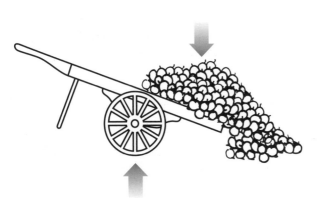

If you moved the wheels forward this would make it more stable, but another problem has been introduced. It is now much more difficult to lift the apple cart ready for moving. In the first design the weight of the apples beyond the position of the wheel helped you lift the handles. In the new design you have to provide all the effort yourself as all the apples are between you and the wheel.

How science helps

You may learn about the properties of materials in science lessons. The **strength** of a material tells you how much force you need to break it. It is important that the material in a structure can bear its load. The load may try to squash the structure or to pull it apart. If the material is not strong enough then parts of the structure will break.

If you know about the strengths of materials you can choose the right one for a particular job. Here is an example.

A thin rope is fine for towing a car but not a lorry; for that you need a steel cable. Although it is the same thickness as the rope it is much stronger.

The **elasticity** of a material tells you how much force you need to stretch, squash or bend it. It is important that the material in a structure is stiff enough to resist the stretching, squashing and bending caused by the load. If the material is not stiff enough then parts of the structure will deflect so much that the structure is unsafe. Here is an example.

A diving board made from polythene would bend so much that the diver couldn't walk to the end. A diving board of the same shape and size made from steel would be so stiff that when the diver jumped on it he would get no spring. The same shape and size of diving board made from pine would have just the right stiffness.

You may learn about **centre of gravity** in science lessons. Objects that are difficult to topple over often have a low centre of gravity. It is easy to push over the person on the left, but when she crouches into a martial arts stance she is much more stable.

Designing structures

Getting an overall design

You can use sketches to develop several overall design possibilities for a structure. These designs are for storage systems to be used mainly for books. You should notice four things:

- they all rely on shelving;
- they all look quite different;
- they all look as though they will work;
- each one can be constructed in several different ways.

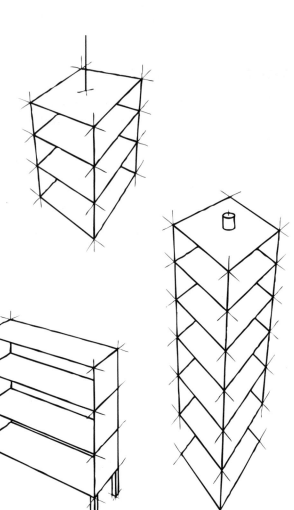

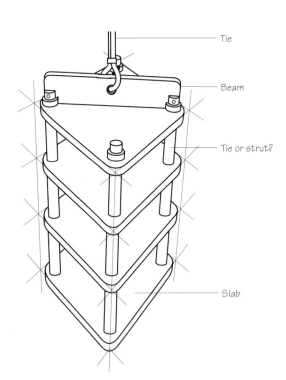

Choosing the parts

Having decided which design idea you like best, you can work out how it might be made up from different sorts of parts (structural elements) and draw a labelled sketch. You can use the Structural Elements Chooser Chart on page 176 to help you.

There are three things to notice about this sketch:

- .the designer has drawn the design very simply;
- she has labelled each piece as a type of part (structural element);
- she hasn't made any detailed decisions yet about material, cross-sectional shape, cross-sectional area or methods of connecting.

Designing each part so that it works well

In this drawing the designer has taken her 'parts only' design and developed it further. There is starting to be enough information for her to make a working drawing for each of the parts and an assembly drawing. Notice that:

- she has specified materials for each part;
- she has specified exact shapes and sizes of each part;
- she has shown exactly how the parts fit together;
- she has shown that the connections between the parts can transfer the loading.

You can use the *Designing points for structural elements* on page 177 to help you produce this kind of drawing.

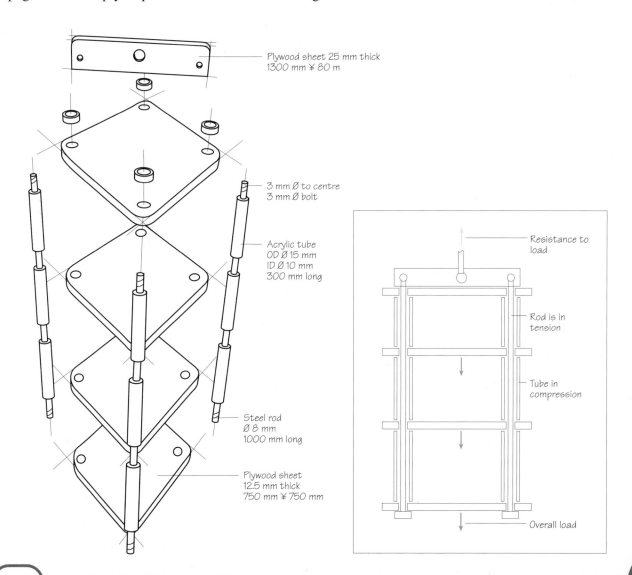

Plywood sheet 25 mm thick
1300 mm ¥ 80 m

3 mm Ø to centre
3 mm Ø bolt

Acrylic tube
OD Ø 15 mm
ID Ø 10 mm
300 mm long

Steel rod
Ø 8 mm
1000 mm long

Plywood sheet
12.5 mm thick
750 mm ¥ 750 mm

Resistance to load

Rod is in tension

Tube in compression

Overall load

Structural Elements Chooser Chart

This chart describes the different structural elements (types of part) you can use in your designs. It explains how each one works. You can use it to decide which elements to use in your design.

Name	Description	How it works
beam	A part that is supported at each end.	The beam resists the load by bending. If the beam is too weak this bending will cause the beam to break in the middle. The beam transfers the weight of the load to the supports at either end. If they cannot resist this load then the structure will collapse.
cantilever	A beam that is held firmly at only one end.	The beam resists the load by bending. If the beam is too weak this bending will cause the beam to break at the support. The beam transfers the weight of the load to the support, pushing both up and down. If the support cannot resist this load then the structure will collapse.
tie	A member in a framework that is in tension. It is being pulled at either end and holds together other members that are trying to move apart.	The tie resists the load by pulling inwards against it. If the tie is too weak the outward pull will be greater than it can resist and the tie will break. If the tie is not stiff enough it will stretch causing the structure to distort.
strut	A member in a framework that is in compression. It is being pushed in at both ends and keeps apart other members that are trying to move together.	The strut resists the load by pushing outwards against it. If the strut is too weak the inward push will be greater than it can resist and the strut will buckle or break.
shaft	A member that transmits turning force (or torque). It is subject to torsion.	The shaft resists the load by twisting against it. If it is not strong in torsion it will break.
hollow box	A 3D cuboid form made from separate sheets joined together. The sides must be able to resist both tension and compression.	The sides of a box are strong in tension but weak in compression. The construction of the box prevents the sides from buckling under the load.
shell	A 3D form made from a single sheet. It must be able to resist both tension and compression.	The load is spread across the whole of the shell. It will be concentrated at sharp-cornered holes and sudden changes of surface so any design should avoid these.

Designing points for structural elements

You can improve the performance of structural elements by modifying their design.

Choice of material

Whichever structural element you are designing you must think carefully about which material to use. A beam made from oak would be both stiffer and stronger than a beam of the same dimensions made from chipboard. A beam made from steel would be over 10 times stronger and about 20 times stiffer than the oak beam but over 10 times heavier!

Beams and cantilevers

Beams and cantilevers can be made:

- *stronger* by increasing their thickness;
- *stiffer* by turning them on their edge;
- *stiffer* by shortening them;
- *stiffer* and *stronger* by changing the cross-sectional shape to an I or a T;
- using less material by removing material from non-critical areas.

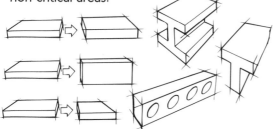

Ties and struts in frameworks

You can make a tie both stronger and stiffer by making it thicker; this increases the cross-sectional area. Making a tie longer will not affect its strength.

You can make a strut stronger and stiffer by making it thicker; this increases the cross-sectional area. Making a strut longer increases

the chances of buckling, making it weaker and more likely to break.

Hollow boxes and shells

You can make a hollow box stronger by ensuring that the sides are firmly attached to each other at the edges and that it has six sides. This will prevent the sides from buckling. You can make a shell less likely to break by using gentle curves for the shape and avoiding rectangular holes. This prevents cracking through stress concentration.

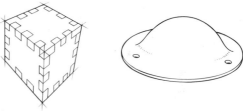

Shafts

You can make a solid shaft stronger and stiffer by making it thicker. If you use a tube for a shaft you must ensure that the walls are thick enough to withstand the twisting forces.

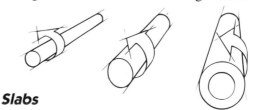

Slabs

You can make slabs stiffer by making them thicker. You can prevent buckling by attaching them to beams.

Tension and compression

The arch bridge and the aerial ropeway in the panels below were designed to resist specific loads and forces. The arch has to resist the load of the tractor. To support the tractor the stones in the arch are squeezed together. This squeeze force is called a **compressive force**, because to resist the load of the tractor the stones in the arch become compressed.

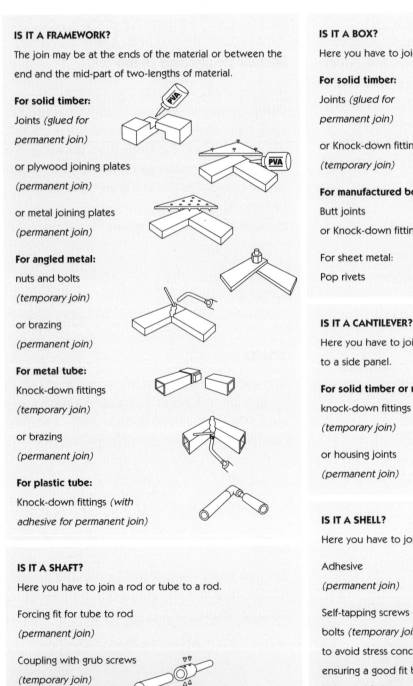

IS IT A FRAMEWORK?

The join may be at the ends of the material or between the end and the mid-part of two-lengths of material.

For solid timber:

Joints (*glued for permanent join*)

or plywood joining plates (*permanent join*)

or metal joining plates (*permanent join*)

For angled metal:

nuts and bolts (*temporary join*)

or brazing (*permanent join*)

For metal tube:

Knock-down fittings (*temporary join*)

or brazing (*permanent join*)

For plastic tube:

Knock-down fittings (*with adhesive for permanent join*)

IS IT A SHAFT?

Here you have to join a rod or tube to a rod.

Forcing fit for tube to rod (*permanent join*)

Coupling with grub screws (*temporary join*)

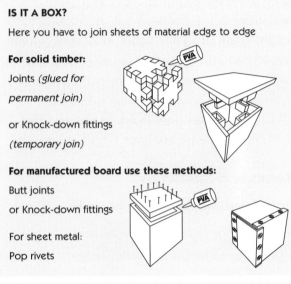

IS IT A BOX?

Here you have to join sheets of material edge to edge

For solid timber:

Joints (*glued for permanent join*)

or Knock-down fittings (*temporary join*)

For manufactured board use these methods:

Butt joints
or Knock-down fittings

For sheet metal:
Pop rivets

IS IT A CANTILEVER?

Here you have to join one end of a wide length of material to a side panel.

For solid timber or manufactured board:

knock-down fittings (*temporary join*)

or housing joints (*permanent join*)

IS IT A SHELL?

Here you have to join the shell to a side panel or a frame.

Adhesive (*permanent join*)

Self-tapping screws or nuts and bolts (*temporary join*), take care to avoid stress concentration by ensuring a good fit between the holes and fittings.

Structures calculations

There will be times when you will need to carry out calculations to work out the details of your design. Here are two examples.

Example 1

Clarissa has designed a bookcase which is hung from a joist in the ceiling. She is concerned about the material she should use for the main tie. This is how she chose which one to use.

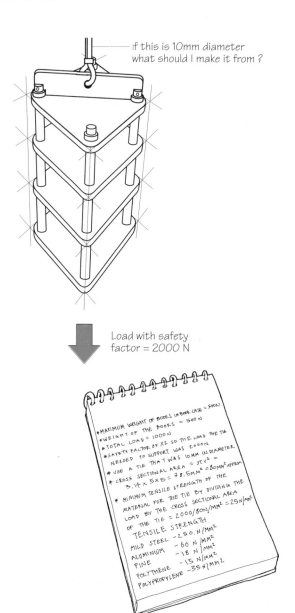

if this is 10mm diameter
what should I make it from?

Load with safety
factor = 2000 N

*MAXIMUM WEIGHT OF BOOKS IN BOOK CASE = 500N
*WEIGHT OF THE BOOKS = 500 N
*TOTAL LOAD = 1000N
*SAFETY FACTOR OF X2 SO THE LOAD THE TIE
NEEDED TO SUPPORT WAS 2000N
* USE A TIE THAT WAS 10MM IN DIAMETER
* CROSS SECTIONAL AREA = JT r²
3.14 x 5 x 5 = 78.5MM² = 80MM² APPROX
* MINIMUM TENSILE STRENGTH OF THE
MATERIAL FOR THE TIE BY DIVIDING THE
LOAD BY THE CROSS SECTIONAL AREA
OF THE TIE = 2000/80N/MM² = 25N/MM²
TENSILE STRENGTH
MILD STEEL - 250 N/MM²
ALUMINIUM - 60 N/MM²
PINE - 18 N/MM²
POLYTHENE - 15 N/MM²
POLYPROPYLENE - 35 N/MM²

Example 2

Nathan has designed a counterbalanced lamp for his study-bedroom. He needs to calculate the weight of the counterbalance. To do this he used the principle of moments.

What else might Nathan need to take into account?

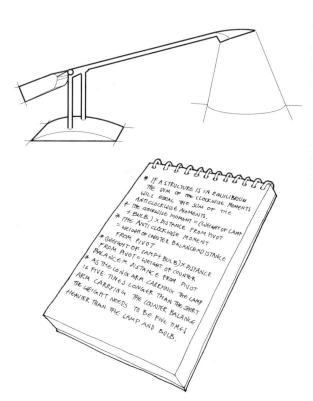

* IF A STRUCTURE IS IN EQUILIBRIUM
THE SUM OF THE CLOCKWISE MOMENTS
WILL EQUAL THE SUM OF THE
ANTICLOCKWISE MOMENTS.
* THE CLOCKWISE MOMENT = (WEIGHT OF LAMP
+ BULB) X DISTANCE FROM PIVOT
* THE ANTI CLOCKWISE MOMENT
= WEIGHT OF COUNTER BALANCE X DISTANCE
FROM PIVOT
* (WEIGHT OF LAMP + BULB) X DISTANCE
FROM PIVOT = WEIGHT OF COUNTER
BALANCE X DISTANCE FROM PIVOT
* AS THE LONG ARM CARRYING THE LAMP
IS FIVE TIMES LONGER THAN THE SHORT
ARM CARRYING THE COUNTER BALANCE
THE WEIGHT NEEDS TO BE FIVE TIMES
HEAVIER THAN THE LAMP AND BULB.

Sometimes you can avoid difficult calculations by using the design decisions of experienced designers. If you are designing seating, for example, you can look at similar products to identify suitable materials and the likely sizes and shapes of different parts.

Revising some basics

The picture shows an exploded view of a toy buggy made to look like a mouse. It is controlled by the switches on the remote controller. These are connected to the buggy by ribbon cable. Below the picture is a circuit diagram showing how the buggy works.

These are the key points to note.

- The LED eye circuit, the buzzer circuit and the motor circuit are in parallel to one another.

- The LEDs come on as soon as the toy is switched on so they act as a 'power on' indicator.

- The LEDs have protecting resistors.

- The on switch is a single pole/single throw toggle switch.

- The eyes can be made to 'blink' with a push-to-break switch.

- The mouse can be made to 'squeak' by operating the buzzer with a push-to-make switch.

- The mouse can be made to go forwards and backwards with a double pole/double throw slide switch with a centre-off position.

- The battery is in the control unit so the toy itself is light and fast moving. A child would have to run after it to keep up.

- The circuit diagram shows how the components are connected together; not how they are laid out in the toy or the controller.

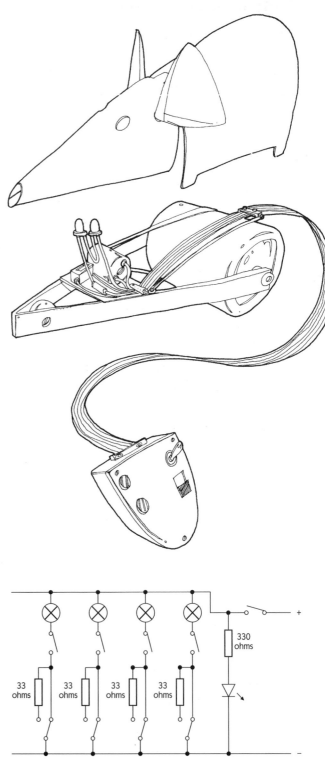

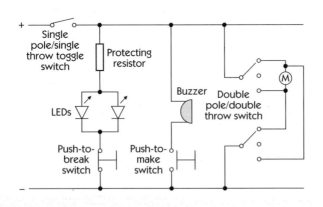

Light sources

For illumination use a lamp or bulb. These may get hot so take care to ensure there is adequate cooling. You can find out about the range of bulbs available on page 186.

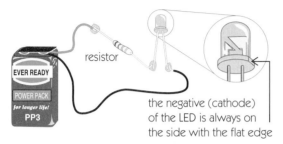

resistor

the negative (cathode) of the LED is always on the side with the flat edge

For indicating, use light-emitting diodes (LEDs). These only give small amounts of light but they do not get hot. As an LED conducts electricity in one direction only it has to be placed in a circuit the correct way round. Some useful information about LEDs is shown in the table.

Type of LED	Shapes and colours available	Needs protecting resistor?	Cost
standard	△ ○ in red, yellow, green	yes	low
high intensity	○ in red, yellow, green	yes	low
ultra bright	○ in red, yellow, green	yes	med
flashing	○ in red	no	high

Micro-switches

A micro-switch is a tiny switch which requires only a small amount of force to operate. This means you can use the moving parts of a mechanical system to operate micro-switches and have automatic electrical control. You can use cams or levers to operate the switches as shown in the example.

Here is the design for a circuit to cause the head of an automaton to turn in a clockwise direction and then in an anticlockwise direction. It uses a timer motor to operate a cam which controls three micro-switches – one push-to-make switch and two single pole/double throw changeover switches. The cam turns slowly because the camshaft is connected to the motor through a compound gear chain which reduces the speed of the motor from 300 r.p.m. to 1 r.p.m.

When motor M2 is switched on nothing appears to happen for 10 seconds until part of the cam operates the push-to-make switch. Then motor M1 comes on, turning the head in a clockwise direction for 25 seconds. Then the cam operates the micro-switches which reverse the direction of the motor and for a further 25 seconds the head turns anticlockwise. The cam then releases all the micro-switches and the procedure repeats.

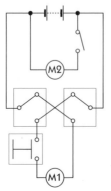

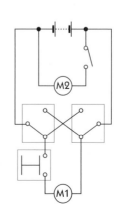

M1 will rotate clockwise   M2 will rotate anticlockwise

Key

☐	Micro-switches controlled by cam
(M1)	Motor operating automaton head
(M2)	Timer motor operating cams
⌐	Switch operating timer motor

Choosing electric motors

You can use the information in the table to help you choose an electric motor suitable for your design.

	Very light duty no gears	Light duty built-in nylon gears	Medium duty built-in steel gears
	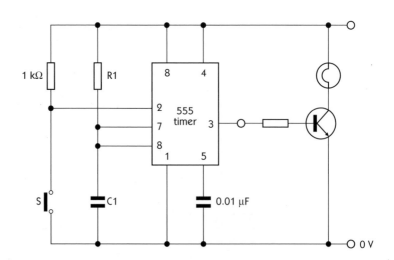		
Size (mm)	20 × 20 × 40	40 × 40 × 80	40 × 40 × 100
Cost	very low	medium	high
Source	Maplins	Radio Spares	Radio Spares
Operating voltage	6–12 V	12 V	12 V
No-load speed		70 r.p.m.	130 r.p.m.
Turning power	very low	medium	high

The monostable circuit

It is sometimes useful in toys or automata to have an electric circuit that operates for a set length of time and then turns itself off. You can use the monostable circuit shown here to achieve this.

It uses a 555 timer microchip (integrated circuit) plus resistors and capacitors to provide the timed output. The circuit also includes a transistor which can provide a larger output current than the microchip. The transistor can be used to operate light bulbs, buzzers and LEDs. For larger current devices it can operate a relay.

The circuit works like this. Before the switch S is pressed the output voltage at pin 3 is low. When the switch S is pressed the voltage at pin 3 goes high and this causes the transistor to switch on and operate either the relay or other output devices.

The time that the transistor stays on depends on the values of resistor $R1$ and capacitor $C1$. It is given by the formula $T = R1 \times C1$ approx., where T = time in seconds, $R1$ = resistance in ohms and $C1$ = capacitance in farads. If $R1$ has a value of 1 MΩ (1 000 000 ohms) and $C1$ a value of 10 µF (10×10^{-6} farads),

$T = 1\,000\,000 \times 10/1\,000\,000 = 10$ seconds.

Changing the value of either $R1$ or $C1$ will change the time the transistor stays on, as shown in the following table.

Output times for monostable circuit		
R1	C1	Monostable output time R1 $\times$ C1 approx.
100 kΩ	1 µF	0.1 seconds
1 MΩ	1 µF	1 second
10 MΩ	1 µF	10 seconds
100 kΩ	10 µF	1 second
1 MΩ	10 µF	10 seconds
10 MΩ	10 µF	100 seconds
100 kΩ	100 µF	1 second
1 MΩ	100 µF	100 seconds
10 MΩ	100 µF	1000 seconds

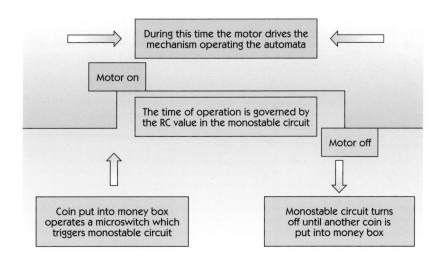

During this time the motor drives the mechanism operating the automata

Motor on

The time of operation is governed by the RC value in the monostable circuit

Motor off

Coin put into money box operates a microswitch which triggers monostable circuit

Monostable circuit turns off until another coin is put into money box

Applying Ohm's law

Sophie wanted to use two lamps in parallel as the eyes in an automaton. She wanted the eyes to get brighter and dimmer. She knew she could use a variable resistor for this and that she could adjust the variable resistor using a crank link and slider system where the slider was a rack operating a pinion on the shaft of the variable resistor. Her problem was choosing a suitable value for the variable resistor. This is how she worked it out:

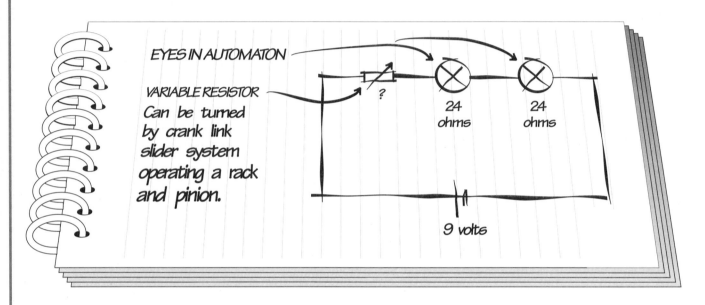

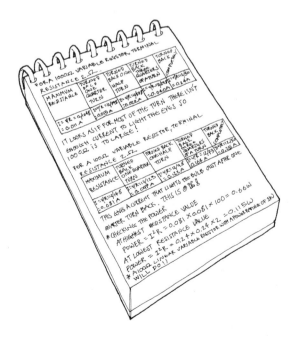

Electrical assembly

Designing a circuit that works is only the first step. You then have to build it so that it works.

The first step is to turn the circuit diagram into a layout drawing and use this to guide your making. The doll's house that Nathan designed is a good example. You can see from the circuit diagram that the light in each room can be turned on or off and be either bright or dim.

In the doll's house

- Lamp holder is fixed into position; not held in place by wiring.

- No bare wire is showing at connection to lamps.

- Wiring is connected to lamps by screw terminals.

- Wiring to individual lamps is laid out neatly and held in place by cable clips.

- Ribbon cable is clamped at back of doll's house.

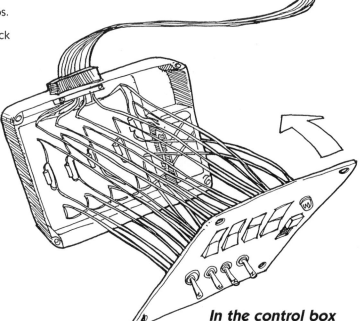

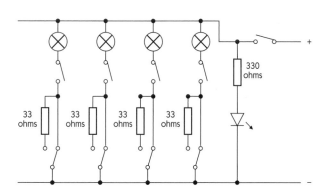

In the control box

- Ribbon cable is used for the wiring from the control box to the doll's house.

- Ribbon cable is clamped at exit from control box.

- Resistors are mounted on matrix board to keep them in fixed position and avoid stress on soldered joints.

- Control box lid has been drilled to take switches which are held in place by screw fittings.

- Battery is held securely in its own compartment.

- Main control switch plus LED show when power is on.

- Power-indicator LED held in position with grommet.

Electrical Components Chooser Chart

You can use this chart to help you decide which components to use in your design.

Batteries suitable for low-voltage direct current devices

non-rechargeable zinc-cabon for low current, infrequent use e.g. torch or calculator	non-rechargeable zinc-chloride for medium current, regular use e.g. hand-held computer game	non-rechargeable alkaline for high current,heavy use e.g. Walkman or radio-controlled model	rechargeable nickel-cadmium for prolonged use e.g. lap-top computer or mobile phone
low cost	medium cost	medium cost	high cost

Cables suitable for internal wiring of electrical and electronic instruments

multistrand copper wire covered in PVC, available in black, blue, brown, green, grey, orange, pink, red, violet, white, yellow	twin figure 8 black, brown, grey, white	ribbon cable covered in grey PVC or colour coded wide variety available e.g. 14-way, 15-way, 16-way, 20-way, etc.
Ensure that wire can carry the required current		
low cost	low cost	high cost

Connectors for joining wires and components

terminal blocks for permanent connection without soldering; take care wiring is not under strain; vibration can loosen the connection	crocodile clips for rapid temporary connection, suitable for test circuits only	solder for permanent connection, take care wiring is not under strain
low cost	medium cost	low cost

Cable fixings used to keep wiring neat and tidy

ties	grips	conduit
low cost	low cost	medium cost

Lamps used to provide illumination —⊗—

Ensure that the lamp matches the power source to avoid burning out. Use the appropriate fitting	
filament	tungsten halogen
low cost	medium cost

Resistors for controlling the size of electric current and potential difference

fixed resistors —▭—	variable resistors —▱—
values shown by the colour code: black 0 brown 1 red 2 orange 3 yellow 4 green 5 blue 6 violet 7 grey 8 white 9	presets adjusted with a screwdriver potentiometers adjusted with fingers
Ensure correct power rating to prevent overheating	
available in a range of preferred values from 10 Ω to 10 MΩ	available in a range of preferred values from 100 Ω to 1 MΩ
low cost	medium cost

Electrical systems

Electrical Components Chooser Chart

Light-emitting diodes (LEDs) for giving small visible signals

non-flashing	flashing
standard, high intensity and ultra bright in red, yellow and green	standard and ultra bright in red
available as △ and O shapes	available as O shape only
needs protective resistor	does not need protective resistor
must be connected correct way round	must be connected correct way round
low cost	medium cost

Motors

very light duty no built-in gears for lightweight applications e.g. moving toys	light duty with built-in nylon gears medium weight applications e.g. automata, rotating signs, point of sale	medium duty with built-in steel gears medium weight applications for prolonged use e.g. testing equipment, pumps
low cost	medium cost	high cost

Noise-makers

buzzers	bells
must be connected correct way round	can be connected either way
must match power source to prevent burning out	
low cost	medium cost

Switches (general)

available as push button, slide, toggle and rocker

to hold something on or off	to set something on or off	to turn something on and something else off	to reverse direction
push-to-make/ push-to-break	single pole/ single throw	single pole/ double throw	double pole/ double throw
low cost	low cost	medium cost	medium cost

Switches (specialist)

micro-switches	available as button, roller or lever	medium cost
tilt switch	operated by vibration or change in position	low cost
reed switch	operated by a magnet	low cost

Metals Chooser Chart

Most metals are obtained from materials taken from the Earth's crust. Metals have some useful properties:

- they are strong, stiff and tough;
- they are dense;
- they are good conductors of electricity and heat;
- they can be polished to show metallic lustres.

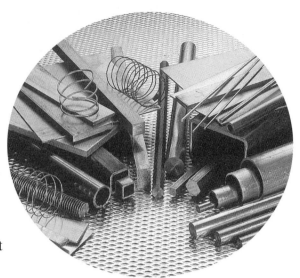

Most pure metals are modified by alloying them with other metals to improve their properties.

Different metals or alloys of metals have different properties. Use the Chooser Chart below to help you choose the best metal for your design.

Metal	aluminium alloy	copper	brass	silver	mild steel	carbon steel
Colour	silver white	pinkisk brown	yellow	bright silver	silver grey	dark grey
Type	non ferrous				ferrous	
Composition	aluminium + traces of copper, magnesium, manganese for hardness	virtually pure copper	alloy of copper and 35% zinc (gilding metal 15% zinc)	silver + 7.5% copper for hardness (sterling silver)	iron + 0.15 – 0.35% carbon	iron + 0.35 – 1.15% carbon
Source	bauxite mined in Africa, Australia and USA	copper, zinc and ore mined in North and South America, Africa and Russia		mined in North and South America	iron ore mined in Sweden, USA and Russia	
Extraction & processing	electrolysis of aluminium oxide produced from bauxite	smelting of ore and electrolysis		smelting	iron produced in a blast furnace is made into steel in a basic oxygen furnace or electric-arc furnace	
Commercial uses	car engine castings & extruded window frames	plumbing and electrical wiring	locks and water taps	jewellery and specialist electrical wiring	car bodies and major structures	tools
Disposal	RC	RC + RU	RC	RC + RU	RC + RU	RC
Durability	tarnishes quickly, corrodes slowly	tarnishes slowly, corrodes very slowly	tarnishes slowly, corrodes very slowly	tarnishes slowly, corrodes very slowly	tarnishes quickly, corrodes quickly	tarnishes quickly, corrodes slowly

Key: RC = commercially recyclable; RU = likely to be found in a form suitable for re-use in school.

Metal	aluminium alloy	copper	brass	silver	mild steel	carbon steel
Melting point/ °C	700–750	1080	950–1000	960	1300–1500	1200–1400
Relative price/£	● x 1.5	● x 10	● x 9 ▲	● x 500 ▲▲	●	● x 2
Ease of sourcing	●●●●	●●●	●●	●	●●●●●	●●●
Hardness (how difficult it is to scratch)	●	●●	●●●	●●	●●●	●●●●●
Strength (how difficult it is to break)	●	●●	●●●	●●	●●●●	●●●●
Density (how heavy it is)	●	●●●●	●●●●	●●●●●	●●●●●	●●●●●
Modulus of elasticity (how difficult it is to stretch)	●	●●●	●●	●●●	●●●●●	●●●●●
Malleability (how easy it is to shape)	●●	●●●	●●●	●●●●●	●	●
Electrical conductivity	●●●●	●●●●	●●●	●●●●●	●	●
School uses	💡	🧲✂	🤖🔧	🧲✂	🤖🔧🪑	🔧
Ease of working by hand	○○○○○	○○○○	○○○	○○○○	○○	○
Ease of machining	○○○	○○○	○○○	○○	○	N/A

Key: ● = the more ●, the greater the property; ▲ = gilding metal (15% zinc and 85% copper) is often used for jewellery (£s x 12); ▲▲ = nickel-silver (18% nickel, 62% copper and 20% zinc) can be used as a cheaper substitute for sterling silver (£s x16); ○ = the more ○, the easier the material is to use ; N/A = not applicable

10 *Plastics Chooser Chart*

Plastics are produced from crude oil, coal and natural gas. They are often referred to as **polymers**. Many are easy to work into different shapes and forms and are available in a wide range of colours and surface finishes. Plastics are resistant to rotting and corrosion. Generally they are stronger than timber but weaker than metals and less stiff than either metal or timber.

There are two main types of plastics.

- **Thermoplastics** will soften when heated and can be shaped and moulded in two and three dimensions.
- **Thermosetting plastics** do not soften on heating. Some are available in sheet form, others as liquid resins which can be made to set.

Thermosetting resins can be reinforced with fibres such as glass or carbon to produce stiffer, stronger materials. These are called **composites**. Solid plastics can be modified to make foams, often called 'expanded' plastics.

The use of modified plastics and liquid resins requires extreme care as dangerous dusts and fumes are often produced. Most plastics are safe provided they are not burnt (which makes toxic fumes) or dry-sanded (which makes a harmful dust).

Different plastics have different properties. Use the Chooser Chart below to help you choose which is the best plastic for your design.

Plastic	acrylic	PVC	nylon	polystyrene	plastic foams	liquid resins	Formica
Name	polymethyl methacrylate	polyvinyl chloride	polyamide	polystyrene	expanded polystyrene, polyurethane foam, polyester foam	polyester resin, epoxy resin	melamine formaldehyde
Abbreviation	PMMA	PVC/UPVC	PA	PS		PR, ER	MF
Type	thermoplastic				modified plastic	thermoset	
Commercial uses	signs	electrical insulators, plumbing fittings	fabrics, combs, bearings	fridge door panels	packing, padding, insulation	boat hulls	work surfaces
Disposal	RU	RC	RC	RC			RU

Key: RC = commercially recyclable; RU = likely to be found in a form suitable for re-use in school.

Plastics Chooser Chart

Plastic	acrylic	PVC	nylon	polystyrene	foams	liquid resins	Formica
Durability	●●●●	●●●●	●●●●●	●●●●●	●●	●●●●	●●●●●
Softening point/°C	85–115	70–80	230	80–105	N/A	N/A	N/A
Relative price/£	●●●	●●	●●	●●	●	●●●●	●●
Ease of sourcing	●●●●	●●●●	●●	●●●●	●●●	●●●	●●●●
Hardness (how difficult it is to scratch)	●●●●	●●	●●●	●●	●	●●●●●	●●●●●
Strength (how difficult it is to break)	●●●	●●●●	●●●●●	●●●●	●	▲	●●
Density (how heavy it is)	●●●	●●	●●●	●●	●	●●●●●	●●●●●
Modulus of elasticity (how difficult it is to stretch)	●●	●●●	●	●	▲▲	▲	●●●●●
School uses	🖼️🖼️	🖼️	🖼️🖼️	🖼️🖼️	🖼️🖼️	🖼️🖼️🖼️	🖼️🖼️
Ease of hand-working	○○○	○○○○	○○○	○○○○○	○○○○	○○	○○○
Ease of processing	○○○○○	○○○○○	○○○	○○○○○	○○○○	○○	○○○
Suitable processes	line bending	vacuum forming	turning component	vacuum forming	hot-wire cutting	casting, moulding	laminating

Key: ● = the more ●, the greater the property; ▲ = dependent on the reinforcing material; ▲▲ = polyester foam is very easy to stretch (value much less than s); ○ = the more ○, the easier the material is to use, N/A = not applicable.

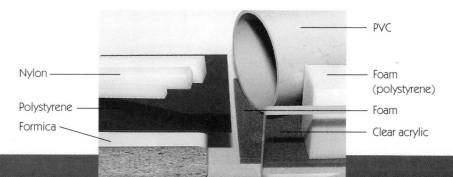

Nylon

Polystyrene

Formica

PVC

Foam (polystyrene)

Foam

Clear acrylic

Timber Chooser Chart

Timber is produced by cutting down trees, sawing them into planks and removing the moisture by **seasoning**. Seasoning requires careful storage and may take many months unless the timber is kiln dried.

Timber can be classified as hardwood or softwood. Hardwoods come from trees which have broad leaves and seeds in fruits or nuts. Softwoods come from trees which have narrow leaves and seeds in cones. This classification tells us nothing about the properties of timber – balsawood is a hardwood which is much lighter and softer than red deal which is a softwood.

Timber is a good structural material but is not as strong, stiff or tough as metal. It is much less dense than metal but has a grain structure which can be featured by polishing. The grain is directional and timber is much stronger and stiffer along the grain than across it.

Timber needs to be treated or finished to prevent rotting and insect attack. It will shrink, swell, distort and crack in different atmospheric conditions depending on whether it is absorbing or losing moisture.

Shakes

Insect attack

Latex canals
(no problem)

Splits

Live knot
(no problem)

Dead knot
(likely to fall out)

Different timbers have different properties. Use the Chooser Chart below to help you choose which timber is best for your design.

Timber	red deal	balsa	mahogany	jelutong	ash	beech	oak
Type	softwood	tropical hardwoods			temperate hardwoods		
Source	Canada, Scandinavia	Central & South America	West Africa & Central America	South-east Asia	America & Europe	Europe	North America & Europe
Commercial use	building, fittings	modelling, boats	furniture	pattern, model making	furniture, sports goods	work-benches	furniture
Disposal	most timbers can be re-used, all are bio-degradable						

Timber	red deal	balsa	mahogany	jelutong	ash	beech	oak
Durability	●●	●	●●●	●●	●●●●●	●●●●●	●●
Relative price/£	●●	●●●●●	●●●●	●●●●●	●●●	●●●	●●●●●
Ease of sourcing	●●●●●	●●●●	●●●	●●	●●	●●	●●
Hardness (how difficult it is to scratch)	●●	●	●●	●●	●●●●	●●●●	●●●●
Strength (how difficult it is to break)	●●	●	●●●●	●●●●	●●●●	●●●●	●●●●
Density (how heavy it is)	●●	●	●●●	●●	●●●●	●●●●	●●●●
Modulus of elasticity (how difficult it is to stretch)	●●●	●	●●●●	●●	●●●●●	●●●●●	●●●●
Stability (how well it keeps its shape)	●●	●●●●	●●●●	●●●●	●●●	●●●	●●●
School uses	🔧🎨🤖🪑	🤖✏️	📦🪑	🔧🤖	📦🪑	📦🪑✏️	📦🪑
Ease of working	○○○○	○○○○○	○○○○	○○○○○	○○○	○○○	○○

Key: ● = the more ●, the greater the value of the property; ○ = the more ○, the easier the timber is to use

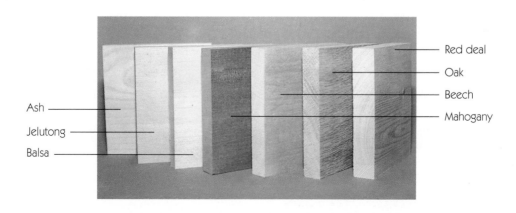

Ash — Jelutong — Balsa — Mahogany — Red deal — Oak — Beech

10 *Modified Timber Chooser Chart*

Solid timber can be used more economically by converting it into modified timber products. By this process sheets of larger size and greater stability can be produced than would otherwise be possible. The solid timber can be cut into small strips and thin sheets called **veneers**. These can be glued together to form plywood or blockboard. It can be broken up into chippings which can be glued and pressed together to form chipboard. If it is ground down into fibres these can be steamed and pressed together to form fibreboards. Most sheets are sold in a standard size of 2440 × 1220 mm, in a range of thicknesses.

By processing more durable types of solid timber and using waterproof synthetic resin adhesives it is possible to make boards which are extremely durable. Exterior grades of plywood and medium density fibreboard (mdf) are used in the building industry and marine plywood is used to make boats. Some sheets of modified timber are sold with a hardwood veneer or plastic coating on the faces. These are especially useful for seating and storage projects. Use the Chooser Chart below to help you choose which is the best one for your design.

Modified timber	plywood	East Asian plywood	marine plywood	blockboard	chipboard	medium density fibreboard	hardboard
Description	birch-faced plywood to a high quality	stout-heart plywood	marine plywood made to a high quality	solid timber core faced with veneers	chippings glued and pressed into a sheet	fibres glued and pressed into a sheet	fibres steamed and pressed into a sheet
Source	Scandinavia	East Asia	East Asia	Scandinavia & East Asia	Central Europe	Britain & Central Europe	Central Europe
Commercial uses	high quality furniture	building trade panels	boat building	high quality furniture	flooring panels	knock down furniture frameworks	packaging & panels on
Durability	●●	●●	●●●●●	●●●	●●	●●	●
Exterior grade	●●●●	●●●●				●●●●	
Relative price/£	●●●●	●●●	●●●	●●	●	●	●
Ease of sourcing	●●●●	●●●●●	●	●●	●●●●●	●●●●	●●●●●
Thickness available	1–25 mm	4–25 mm	4–25 mm	12–25 mm	3–25 mm	2–50 mm	3–6 mm
Strength	●●●	●●	●●●	●●●	●●●●	●●●●●	●●●
Density	●●●●	●●●	●●●●●	●●●●	●●	●●●	●
School uses		large panels & bases	boat building		large panels & bases		stage sets

Key: ● = the more ●, the greater the value of the property

Working with resistant materials

When working with resistant materials you will need to use abrasives, adhesives, fixings, fittings and finishes. The Chooser Chart below summarizes commonly available options.

	Metals	Plastics	Solid and modified timber
Abrasives	Different grades available, begin with coarse and work down to fine emery cloth aluminium oxide paper with a little oil water of ayr stone used wet metal polish	silicon carbide paper (wet and dry paper) used wet with water 000 steel wool used dry with water metal polish	glass paper aluminium oxide paper used dry with a cork block flour paper
Adhesives	Read and follow instructions in case of hazardous substances epoxy resin (Araldite) cyanoacrylate (superglue)	epoxy resin (Araldite) cyanoacrylate (superglue) contact adhesive (Thixofix)	epoxy resin (spa-bond for wood) cyanoacrylate (superglue) contact adhesive (Thixofix) synthetic resin (Resin W, Cascamite & Aerolite)
Heat joining	Always follow safety procedures when using heat joining methods soft solder hard solder & flux brazing & flux arc welding gas welding MIG welding		
Fittings and fixings	bolt machine screw or setscrew countersunk head pan head self-tapping screw rivet blind rivet nut wing nut washer cheese head round head		woodscrew countersunk head raised head round head slotted cross point nails and pins oval French wire panel pin veneer pin hinge knock down fittings
Finishes	Always follow safety procedures. Read and follow instructions in case of hazardous substances metal polish buffing paint lacquer plastic dip-coating enamel (for copper) anodizing (for aluminium) oil blueing (for steel) buffing	metal polish buffing	paint varnish sanding sealer coloured stain french polish wax polish

This section describes the techniques you may need to make your product. You can use the information in this section to choose the techniques that you need to make your design. You will need to get detailed advice and guidance from your teacher on the safest ways to carry out these techniques.

Marking out

Good making starts with careful marking out. Since most products are rectangular or geometric in shape, the easiest and most accurate method of marking out your work is to work from a straight edge which you have checked or created. You can draw lines at right-angles to this edge using a try-square for wood and an engineer's square for metal and plastics. Draw lines parallel to this edge with a marking gauge or odd-leg callipers.

Draw curves or circles on wood with a sharp pencil and a compass, but on metal or 'bare' plastic you must 'punch' the centre to locate a pair of spring dividers securely. For all marking use a pencil for wood and a scriber or marker for plastics and metal.

Precision marking out

Use a cast-iron surface plate which has been machined perfectly flat so that it forms a datum for all measurements on the work being marked out. Then clamp the work to an angle plate which has also been accurately machined so that, even on its side, it lies at 90° to the surface plate.

Use a surface gauge, which can be finely adjusted against a ruler, to transfer measurements to the workpiece. Clamp cylindrical work in a vee block.

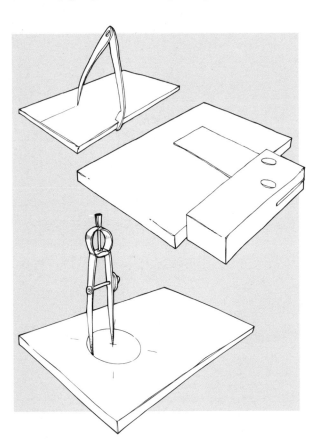

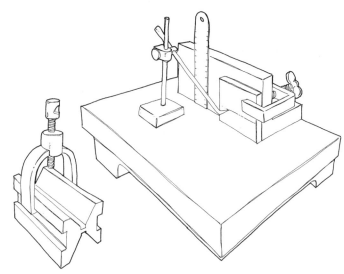

Making box structures

From a single sheet of thermoplastic

1 Make a card model of your box.

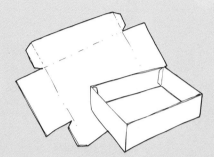

2 Mark out the development on the thermoplastic sheet. Use datum edges at 90° to ensure accuracy.

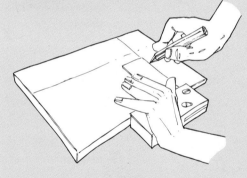

3 Carefully cut out the required shape.

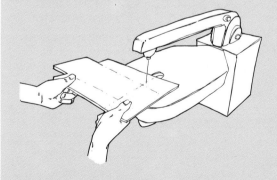

4 Finish edges by draw-filing and then use wet-and-dry paper to achieve a smooth finish. This can then be polished by hand or with a buffing machine.

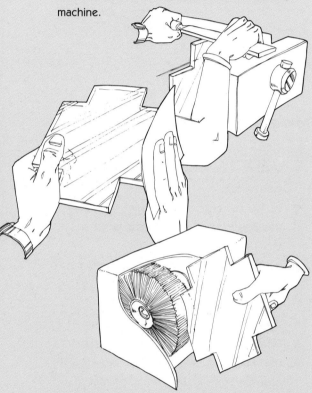

5 Mark the position of the folds with a Chinagraph pencil. Use a strip heater to soften the plastic along the required lines before folding to the correct angle.

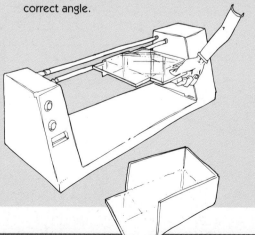

From separate pieces of thermoplastic

You can construct a box by assembling separate parts using adhesives such as Tensol, and mechanical fixings such as blind rivets or self-tapping screws.

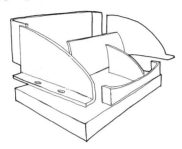

From separate pieces of wood

It is important to prepare the starting material so that it is the correct size with edges and faces planed square. There are several ways to assemble a box from pieces of wood.

From a single sheet of metal

1 Coat sheet metal with marking blue before marking out.

2 Use file and engineer's try-square to establish true edges at right-angles. These edges are then used as datum lines.

3 Use oddleg callipers to mark lines parallel to these true edges. Mark other lines with a scriber.

4 Use tin snips to cut out the development.

5 File at edges to achieve an accurate finish.

6 Use folding bars held in a vice to provide a 90° edge for bending. Use a rawhide mallet to bend by hammering gradually up and down the length.

7 Mark and drill holes before pop-riveting.

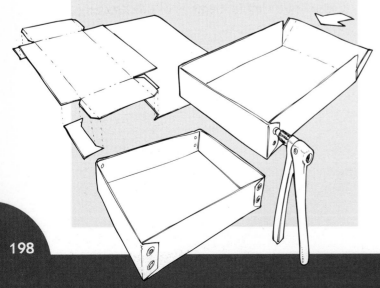

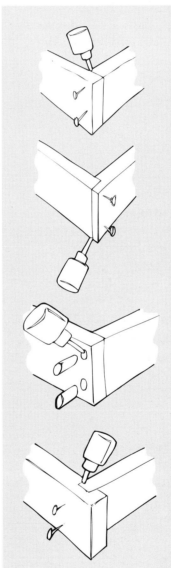

Butt joint
Simple to make but little strength as it relies on glue and nails.

Lap joint
Slightly stronger than a butt joint as the 'shoulder' gives rigidity but still relies on glue and nails for strength.

Dowel joint
Stronger than lap or butt joints as dowel pins hold the separate parts together.

Housing joint
Slightly stronger than a lap joint; often used for partitions.

Comb joint
This has greater strength because the pieces inter-penetrate. It is difficult to make and takes a long time.

Making frame structures

The advantage of a frame structure is that it combines strength with low weight.

Wooden frames

A range of methods of building *flat* frames is illustrated below. All these joints require glue and should be clamped while drying if possible.

You can construct 3D frames by joining flat frames together.

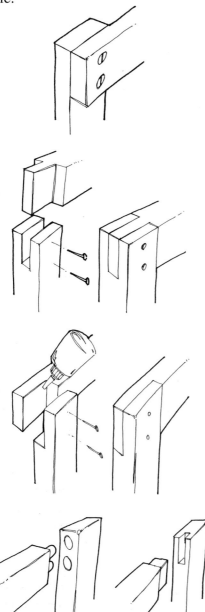

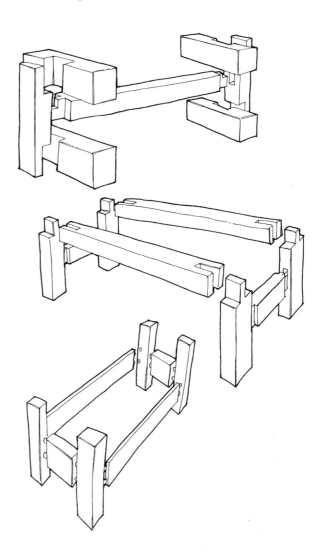

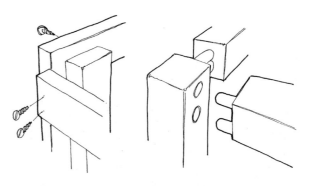

Here are two simple methods of constructing corners in 3D frames.

Metal frames

You can construct frames by joining pieces of tubing. The tubing can be brazed together or joined mechanically.

Brazing

1 Clean joint with file or emery cloth.

2 Mix flux with water to a creamy consistency and apply to joint.

3 Secure joint in position with wire or with firebricks.

4 Heat joint until both pieces turn a very bright red.

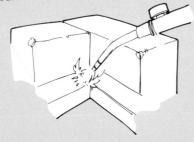

5 Touch the brazing spelter onto the joint. The heat will melt the spelter and it will run along the joint.

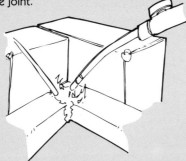

6 After cooling any excess spelter may be removed by filing.

Mechanical methods

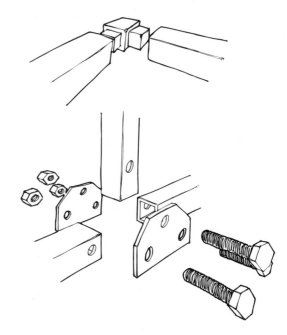

You can also construct frames by bending metal rods. You need to use a bending jig, as shown. This is particularly useful if you have to bend several rods to the same shape.

If you need to bend metal tube then you should use a tube bender to prevent the metal from buckling.

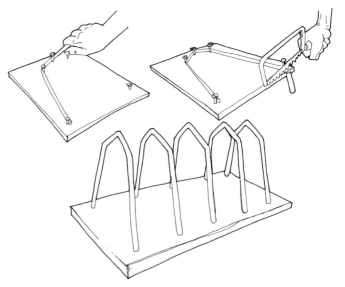

▶ *Using a bending jig for metal rod*

Making shell structures

Hollowed, concave, convex and other three-dimensional forms can be created in wood, metal and plastics. Some methods are outlined here.

Shell forms from sheet metal

You can make hollow dish shapes from copper, brass, guilding metal or aluminium.

1. Cut out the shape with tin snips.
2. Smooth the edges by draw-filing.
3. Hollow it by beating with a bossing mallet on a sandbag. Work in concentric circles starting at the edge and moving towards the centre.

4. Anneal the work at intervals by heating to red heat and plunging into cold water.

5. Use a round-headed stake to produce tall and steep-sided shell forms.

Shell forms from laminated wood

Simple bent shapes can be formed by gluing together thin strips of wood (veneer) and making parallel saw cuts.

More complex and elegant curved shapes can be formed by gluing veneers together under pressure over a former.

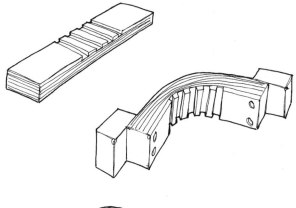

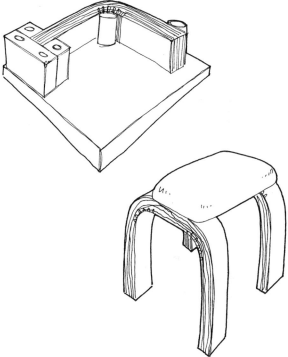

Sheets of thin plywood or hardboard can be bent around a solid framework to form curved structures.

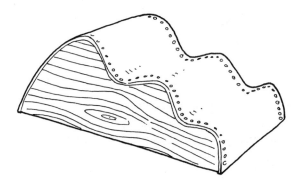

Bending thermoplastics

You can bend acrylic plastic sheet by using a strip heater. To make more complex shapes or 'repeats' you will need to use a bending jig.

You can bend whole sheets of acrylic by heating in an oven and then draping over a former and leaving to cool.

Using a strip heater

1 Heat until soft enough to bend.
2 Use bending jig to get the required angle.

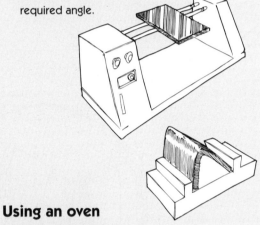

Using an oven

1 Heat in an oven.
2 Drape softened acrylic over a former.

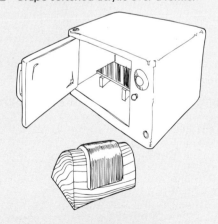

Forming thermoplastics

You can form thermoplastic sheet in two ways:

● plug-and-yoke forming;
● vacuum forming.

Plug-and-yoke forming

1 Cut out plug shape from mdf.
2 Glue yoke onto a basepiece.
3 Place hot, soft plastic sheet over yoke.
4 Press plug into yoke.
5 When cool remove formed plastic sheet.
6 For thicker sheet use a smaller tapered plug and clamps to apply pressure.

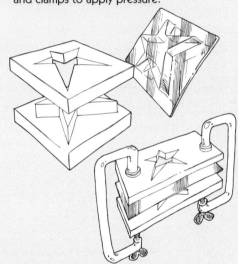

Vacuum forming

1 Clamp thermoplastic sheet into position and heat until soft.
2 Place it over the former and switch on the vacuum pump.
3 Removing the air causes atmospheric pressure to force the soft plastic over the former.
   Note: you can use card and string formers for low-relief forms.

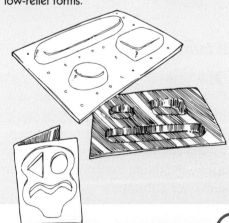

LIRT4

Making shapes using hand-tools

You can shape a piece of material by cutting away or **wasting** the unwanted parts. The remaining piece can then be trimmed to size and finished. The tools you can use for this are described in the Hand-Tool Chooser Chart.

	Wood	Metal	Plastics
Straight lines and flat surfaces	for cutting to length for cutting sheet for trimming	for cutting to length for cutting sheet for trimming	for cutting to length for trimming
Curved lines	for cutting for trimming	for cutting for fine work for trimming	for cutting for fine work for trimming
Curved surfaces	for cutting for trimming	for cutting to approx shape for trimming	for cutting to approx shape for trimming
Round holes	for marking for holes up to 6 mm for larger holes	for marking for holes up to 6 mm for trimming	for marking for holes up to 6 mm
Square holes	for through holes in thin material for flat-bottomed holes	for through holes in thin material for trimming	Drill 3 mm hole inside required shape and use coping saw as for wood.
Irregular-shaped holes		for through holes in thin material for chain drilling for trimming	
Grooves		Use a milling machine.	Use a milling machine.
Slots		for chain drilling for trimming	for chain drilling for trimming

Making shapes using machine tools

You can cut away material using machine tools powered by electric motors. You may have already used an electric hand drill and pillar drill. Here are some more examples.

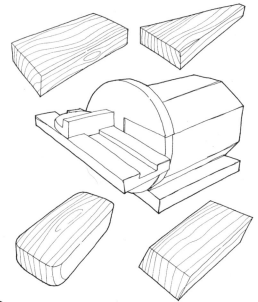

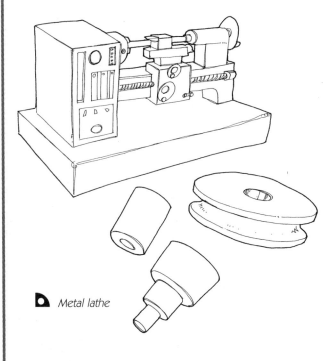

▶ Sanding wheel

▶ Metal lathe

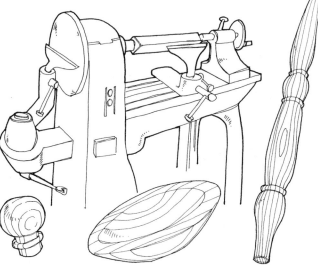

▶ Wood turning lathe

Computer-assisted manufacture

Both milling machines and centre lathes can be computer controlled. This enables the manufacture of complex parts and batch production (see page 82 for examples).

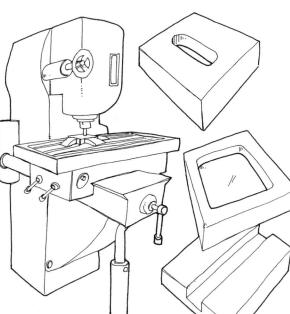

▶ Milling machine

Making moving parts

Sometimes your design will need parts that are attached but able to move. Often you can buy components for this purpose, such as hinges.

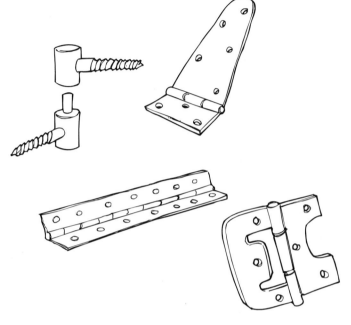

There may be times when you will need to design and make your own methods of attachment. Here are some examples.

For sliding movements:

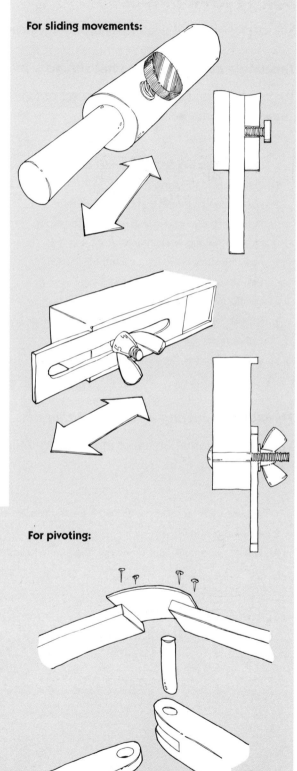

For simple hinges:

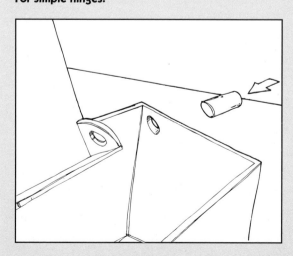

For pivoting:

Making screw threads

Sometimes you will need to make screw threads for your design.

Tapping – cutting an internal thread

To cut an internal thread you use a set of taps which cut the screw thread inside an already drilled hole.

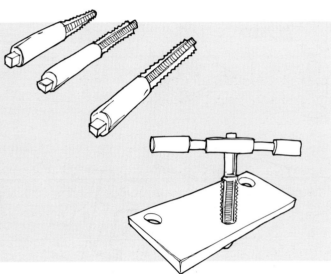

1. Use the chart to select the correct drill size.
2. Drill the hole for tapping.
3. For all sizes of thread there is a set of three taps which are used in sequence. Fit the taper tap into the tap wrench and locate it in the hole. Turn clockwise to cut the thread. Each turn should be followed by a half turn anticlockwise to release the swarf.
4. Repeat the procedure with the second tap and the plug tap. *Note:* on thin materials it is only necessary to use the plug tap.

Threading – cutting an external thread

You cut an external or screw thread using a tool known as a **die** which is held in a die stock.

1. Choose the exact size die to match the bar you are cutting, i.e. M8 for 8 mm bar.
2. Taper end of bar.
3. Locate die on bar. Turn clockwise one turn followed by a half turn anticlockwise.

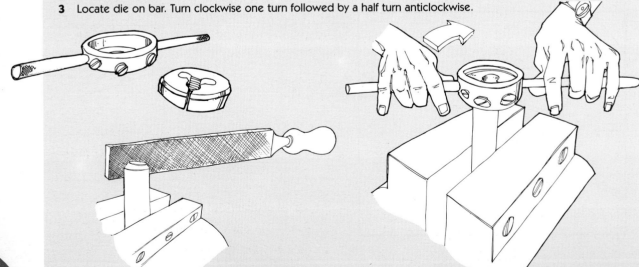

Special techniques for body adornment

Casting

Metal casting in schools is commonly restricted to aluminium and pewter (an alloy of tin and lead). These can be melted at fairly low temperatures and are comparatively cheap.

You can use **cuttlefish casting** for simple 3D forms and **lost wax casting** for complex 3D forms.

Cuttlefish casting

1 Saw cuttlefish bone in half and flatten both cut surfaces.
2 Push locating pegs through both pieces.
3 Place model between both halves and press together, or cut out design.
4 Cut out pouring hole and vent holes.
5 Pour molten metal into cuttlefish mould.
6 When metal has solidified split mould and clean up.

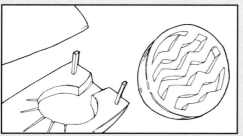

Sheet cutting and piercing

You can cut intricate designs from sheet metals such as copper, brass and aluminium using a piercing saw.

1 Transfer your design to the metal sheet. A self-adhesive label with the design drawn on it is usually suitable. For very fine work use engineer's blue and a scriber.
2 Ensure that the saw is fitted with a blade suitable for the thickness of the metal you wish to cut (check with your teacher).
3 Saw on a bench pin. Keep moving the work so that the part you are cutting is always over the gap in the bench pin.
4 Use a gentle, steady up and down movement to saw, allowing the saw to do the work.
5 To cut internal shapes, drill a hole near to the cutting line, release the saw blade and thread it through the hole.

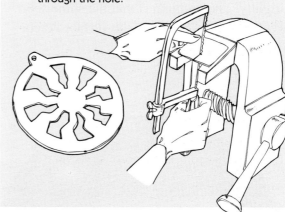

Forming wire

Use simple jigs made from dowel or nails driven into a piece of hardwood for repeated shapes that will be soldered together.

You can use flattened wire for decoration or to make a pin 'catch'. You can make wire flat by annealing and hammering.

Quality finishes

Why finish?

Finishes are applied for one or more of the following reasons:

- to change the appearance of the material;
- to protect the material from dirt, damage, corrosion or decay;
- for easy maintenance by dusting, washing and polishing.

Here are some examples of finishes that you can use on the products that you design and make.

Finishing metal

The amount of shine from a metal surface depends on how smooth the surface is. A smooth, highly polished metal shines brightly. The problem is that the shine soon wears off most metals as the surface reacts with the atmosphere.

▷ *The Hammerite paint on this metal tool box provides a scratch-resistant surface which also prevents the steel from rusting. It is easy to apply and comes in a smooth or dimpled finish*

▷ *This rack is made from steel wire and it has been dip-coated. The white polythene coating looks attractive, prevents the steel from rusting and is much quicker to apply than paint*

▷ *These nobium earrings have been anodized. The lustrous colours are thin films of metal oxide which protect the metal from reaction with the atmosphere*

▷ *These copper brooches have been decorated with enamels. They provide a much more durable finish than paint.*

Finishing wood

To obtain a quality finish on wood, the surface should be as smooth as possible. This is when the grain looks at its best and any surface coating goes on without blemish. Dents, scratches and glue spots only look worse when painted over!

This garden chair has been treated with a microporous varnish. This has tiny pores which are too small to allow rainwater beneath the surface but large enough to allow moisture to escape from the wood and prevent blistering

This bowl has been stained so that the colour is altered but the grain pattern is still visible. The stain does not protect the wood from rotting

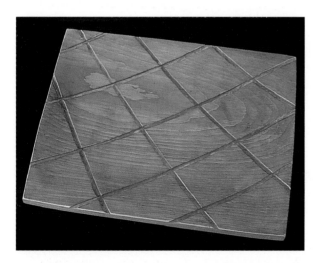

The natural beauty of the wood grain has been enhanced by using a wax polish on this wooden fish

Finishing plastics

A quality finish on plastic depends on the surface being as smooth as possible so that it can be polished. Scrapes, scratches and glue spots are impossible to hide.

The body of this 'Save the Whale' badge is made from acrylic plastic. Both the surface and the edges are smooth and well polished. The 'Save me!' lettering is made from adhesive-backed PVC. Neither the PVC nor the acrylic need any further finishing as they are not attacked by the atmosphere.

Acrylic badge

Finishes and safety

Finishing processes often involve hazards. Use the panel for advice on how best to control the risks.

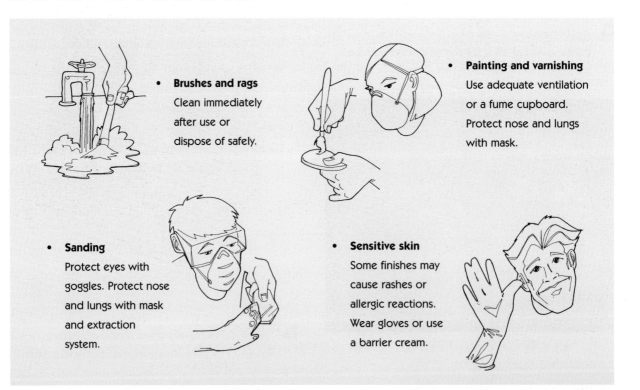

- **Brushes and rags**
  Clean immediately after use or dispose of safely.

- **Painting and varnishing**
  Use adequate ventilation or a fume cupboard. Protect nose and lungs with mask.

- **Sanding**
  Protect eyes with goggles. Protect nose and lungs with mask and extraction system.

- **Sensitive skin**
  Some finishes may cause rashes or allergic reactions. Wear gloves or use a barrier cream.

Important ideas

To avoid accidents and harm in a situation, you need to think about:

- hazards;
- risks;
- risk assessment;
- risk control.

A *hazard* is anything which might cause harm or damage. The chance of a hazard causing harm or damage is called the *risk*. You can work out how big the risk is by thinking about whether the harm or damage is likely to happen. This is called *risk assessment*. *Risk control* is the action taken to ensure that the harm or damage is less likely to happen.

Looking at an unfamiliar situation

The picture shows a small factory which produces stools. Some of the activities have been labelled. Think about the hazards and risks associated with each activity.

1	Delivery of timber sheets
2	Delivery of metal tubes
3	Storage of timber sheets
4	Storage of metal tubes
5	Delivery of fixables and consumables
6	Moving timber sheet
7	Moving metal tubes
8	Cutting sheet into seat shapes
9	Cutting tube to length for frames and foot-rests
10	Painting seats
11	Bending tubes to shape for frames and foot-rests
12	Welding on foot-rests

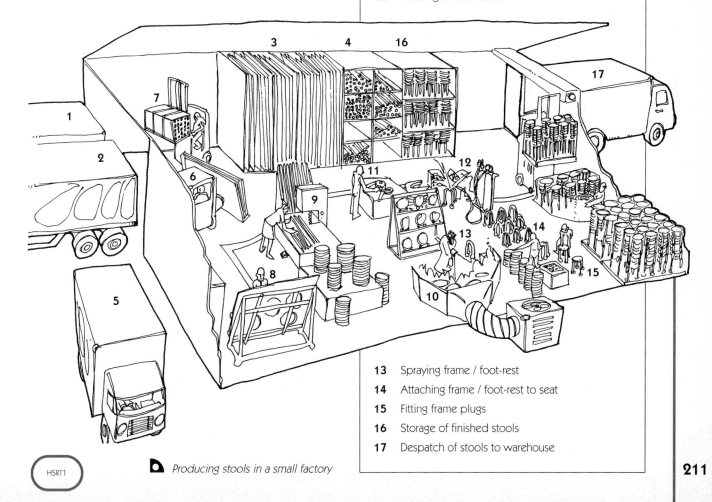

13	Spraying frame / foot-rest
14	Attaching frame / foot-rest to seat
15	Fitting frame plugs
16	Storage of finished stools
17	Despatch of stools to warehouse

Producing stools in a small factory

This table describes the factory's production process and for each there are notes showing hazard identification, risk assessment and risk control. It is important that you learn to look at situations in these terms so that you can take appropriate risk control actions.

You can use this approach to avoid accidents and harm anywhere – at home, at school, at work and when travelling – and whatever you are doing.

Activity	hazards	Risk assessment	Risk control
delivery of timber sheets	strain injuries	high	staff training, working in pairs
delivery of metal tubes	strain injuries	high	staff training, working in pairs
storage of timber sheets	strain injuries	high	staff training, working in pairs
storage of metal tubes	strain injuries	high	staff training, working in pairs
delivery of fixings and consumables	strain injuries	low	staff training, working in pairs
moving timber sheet	strain injuries	high	staff training, working in pairs
moving metal tubes	strain injuries	high	staff training, working in pairs
cutting sheet into seat shapes	strain injuries, major cutting injuries, toxic injuries, sight injuries	high	staff training, machine guards, goggles, face masks, dust extraction
cutting tube to length for frames and foot-rests	strain injuries, major cutting injuries, sight injuries	high	staff training, machine guards
painting seats	toxic injuries, sight injuries	high	staff training, face masks
bending tubes to shape for frames and foot-rests	strain injuries, minor cuts and bruises	medium	staff training
welding on foot-rests	burn injuries, sight injuries	high	staff training, face masks
spraying frame/foot-rest	toxic injuries, sight injuries	high	staff training, face masks
attaching frame/ foot-rest to seat	sight injuries, minor cuts and bruises	medium	staff training, goggles
fitting frame plugs	sight injuries, minor cuts and bruises	low	staff training, goggles
storage of finished stools	strain injuries	high	staff training, working in pairs
dispatch of stools to warehouse	strain injuries	high	staff training, working in pairs

Safe to use?

Bicycles are used by many people all over the world. They are safe only if used properly. Various products and information have been developed to encourage people to do this. In the UK the law states that bicycles ridden at night must carry front and rear lights and a red rear reflector. At the moment it is recommended, though not compulsory, that cyclists wear a cycle helmet and clothing that can be seen easily.

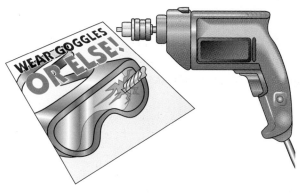

It is much more difficult to ensure that the use of a tool, such as an electric drill, is safe. It has to:

● be produced to the required safety standards;

● have no electrical or mechanical faults;

● be convenient to use;

● work well for its intended purpose.

However, accidents still occur and people get hurt. What can be done to prevent such accidents or at least minimize the harm caused?

The main way to prevent accidents is for the people using the tools to be aware of the hazards and to take appropriate risk control action. In the case of an electric drill this would involve:

● ensuring that the twist drill is firmly and accurately located in the chuck;

● removing the chuck key before drilling;

● switching off the power when changing drills;

● not using blunt drills;

● wearing safety goggles;

● wearing a face mask if dust is produced.

In the workplace employees are obliged by law to follow codes of safe practice. At school you are expected to behave responsibly in a way that will control risk. The consequences of failing to take risk control action can be severe.

◗ *Products, information and legislation enable you to control some of the risks in cycling*

Information for safety

Signs in the workplace

Many places and activities are dangerous. Usually there are signs giving information or instructions to help control the risk. For example, in all workshops there are signs indicating that eye protection should be worn when operating machines; sometimes there are signs indicating the need for ear protection. A selection of these signs is shown below.

Codes of practice

All industries and businesses are required by law to meet the requirements of the Health and Safety at Work Act. Under this legislation it is the responsibility of the employer to provide a working environment in which the risks have been assessed and are controlled by safe working practices.

The employer has a duty to ensure that employees receive appropriate training and instruction in safe working practices. It is the responsibility of employees to work in a way that follows safe working practice. If the employer fails to meet these obligations then the company can be prosecuted, fined and forced to meet the requirements of the Health and Safety at Work Act. If an employee fails to meet his/her obligations, they face warnings, disciplinary procedures and dismissal.

Long-term dangers

One of the most difficult areas of health and safety is to assess the risk associated with long-term exposure to low levels of potentially harmful substances. Occasional exposure to dust, fumes, finishes, cleaning materials, etc. may not be harmful, but regular exposure, every day over a working life may cause irreparable damage and reduce quality of life. There are scientists who specialize in investigating these long-term effects and their findings and recommendations are playing an increasingly large part in risk identification and assessment.

 A selection of safety signs

Glossary

3-bar linkage : solid linkage system made from 3 links used in adjustable structures

4-bar linkage : non rigid linkage system made from 4 linkages used in mechanisms

angular momentum : the property of a spinning object that keeps it pointing in the same direction

anodising : the use of electrolysis to build up a protective oxide film on a metal

anthropometrics : design considerations taking people's sizes and shapes into account

approximate values : values that are not exact but of the correct order of magnitude; for example if a switch broke after being used between 1500 and 2000 times you could say this switch can be used for approximately 1750 on-off operations

automata : mechanical toys that perform a series of movements automatically

brainstorming : a process for generating design ideas

cam : a non circular wheel that rotates and moves a follower. It can convert rotary movement into oscillating or reciprocating movement

capability tasks : designing, making and testing a product that works

case studies : real examples of design and technology

cellular manufacturing : manufacturing system using small teams of people called cells

centre of gravity : the point in an object through which the weight of the object acts

centrifugal governor : a means to control the speed of a rotating shaft

closed design brief : a summary of the aims of a design indicating the kind of product that is needed in sufficient detail to limit the range of solutions possible

comparison : a method of testing one product against another

composites : materials made from different materials joined together resulting in a material with improved qualities. The use of glass fibre with thermosetting resins increases both the stiffness and strength of the resulting material

compression : a squeezing force

computer-aided design (CAD) : designing using computers to produce 2D or 3D drawings or surface envelopes or simulations of electrical and mechanical systems

computer assisted manufacture (CAM) : manufacturing process which is controlled by computers

cuttlefish casting : metal casting using mould made from two pieces of cuttlefish bone

descriptions : a phrase or sentence summarising the appearance of an object

die : tool for cutting a screw thread

dynamometer : testing device which measures power

elasticity : a measure of the stiffness of a material; very elastic materials are easy to stretch

electroplating : the laying down of a thin film of metal onto another metal by means of electrolysis

end float : the space between an object rotating on a shaft and the chassis supporting the shaft

ergonomics : design considerations regarding people's movements

exploded view : illustration which shows how the different parts of a product fit together by showing the product pulled apart

feedback : information sensed by closed loop control systems so that they can respond to change

findings : components for making body adornment; usually bought in rather than made

first angle projection : a system of orthographic drawing used for plans or working drawings

fittings : components such as hinges and handles

fluorescent light : a lighting device in which a visible light is produced by the action of invisible UV light on phosphorescent material

gear : toothed wheel usually fixed to a shaft so that it rotates with the shaft

gear train : gear wheels with teeth meshed together so that one drives the other

Geneva mechanism : a mechanisms which converts continuous rotary movement to intermittent rotary movement

grouped results : a collection of numerical results within defined limits

image board : a collection of pictures of things that a person or group of people might like, places they might go or activities they might do

incandescent light bulb : light bulb producing light by means of a tungsten filament which glows white hot when electricity passes through it

injection moulding : a manufacturing process for producing parts with complex 3D forms; molten plastic is forced into a metal block containing cavities which are the form of the parts required

interfaces : links between different systems

interviewing : finding out about people's preferences by asking questions and listening to the answers

isometric drawing : a system of drawing objects in 3D; lines of equal length along the axes in the object being drawn appear as such in the drawing

just-in-time manufacturing : manufacturing system organised to meet immediate market demand. No large stocks of materials, components or finished products are held

lever : a bar or rod that applies a force by moving about a pivot or fulcrum

line of interest : a term used to define a group of similar products eg seating or lighting

lost wax casting : a manufacturing process for producing body adornment parts with complex 3D forms. A plaster mould is formed around a complex 3D wax form. On heating the wax melts and is poured away leaving a cavity. Molten metal is poured into this cavity to give the form required

tungsten halogen bulb : a tungsten filament light bulb containing halogens which prevent breakdown of the filament by forming volatile tungsten halogen compounds which re deposit tungsten onto the filament once the bulb is switched off; available as mains and low voltage

numerical values : observations that give results in the form of measurements provide numerical values; i.e. specific numbers

open design brief : a summary of the aims of a design offering the possibility for a wide range of alternative solutions

orthographic projection : a way of drawing the detail of a 3D object in 2D by showing square on views of different sides

performance specification : a description of what a product should do, look like and any other requirements it must meet

phenakistoscope : optical toy using animation

plans : drawings of a product that give the information needed to make it (sometimes called working drawings)

polymer : the general name given to substances composed of many small molecules joined into long chain molecules; all plastics are polymers.

product design : the area of designing and making concerned with products made from resistant materials and technical components

render : techniques to make the surface textures on drawings look more realistic

resource tasks : short, practical activities to develop knowledge and skills

reviewing : checking your design ideas as they develop against the specification

seasoning : removing the moisture from timber by slow drying

Sick Building Syndrome (SBS) : the illness experienced by people who work in buildings which have been poorly constructed and suffer from poor air conditioning and lighting

single point perspective : a system of drawing in which all horizontal lines converge at a single vanishing point; useful for interiors

strength : a measure of the force required to break a material

tensile strength : the pulling force required to break a material

tension : a stretching force

thaumatrope : an optical toy which uses persistence of vision to see an image formed by a rapidly spinning disc

thermoplastics : plastics which soften reversibly on heating and can therefore be shaped and formed by the use of heat.

thermosetting plastics : plastics which do not soften on heating

third angle projection : a system of orthographic drawing used for plans or working drawings

three point perspective : a system of drawing with three vanishing points; gives an aerial view

two point perspective : a system of drawing with two vanishing points

veneers : thin sheets of timber that are used to make laminates eg plywood and to decorate the surface of manufactured boards

wasting : shaping a material by removing the unwanted parts

working drawings : drawings of a product that give the information needed to make it (sometimes called plans)

zoetrope : an optical toy in which images on the inside of a rotating drum are viewed through slits in the drum and appear to move

Index